Reinforced Concrete Design I
Lecture Notes

A Student Practice Textbook for Structural Engineering

Peter I. Kattan

Petra Books
www.PetraBooks.com

Peter I. Kattan, PhD

Correspondence about this book may be sent to the author at one of the following two email addresses:

pkattan@petrabooks.com

info@petrabooks.com

Preface

These are the handwritten notes for the course "Reinforced Concrete Design I" that was taught at Applied Science University in the period 2000-2002. These notes are based on the book "Reinforced Concrete Design" by Wang and Salmon, Third Edition. This book is out of print at the present time. Sutdents find these notes useful and helpful in their studies and the author is glad to make them available to students worldwide. It is hoped that these notes will also revive the Wang and Salmon book.

The notes are freely available on the internet for free download as individual chapters. But their existence in one single volume is available in this book format only.

February 2024   Peter I. Kattan

Reinforced Concrete Design I Lecture Notes
A Student Practice Textbook for Structural Engineering

Contents

Reinforced Concrete Design I Lecture Notes
A Student Practice Textbook for Structural Engineering

<u>Reinforced Concrete Design (1)</u>

[1

<u>Chapter 1</u>
<u>Introduction</u>

1.1 Definitions:

* <u>Concrete</u> is a mixture of sand, gravel, crushed rock, or other aggregates held together in a rocklike mass with a paste of cement and water.

* Concrete has a high compressive strength and a <u>very</u> low tensile strength.

<u>Types of Structural Concrete:</u>

(1) Plain Concrete.
(2) Reinforced Concrete.
(3) Prestressed Concrete.

* <u>Reinforced Concrete</u> is a combination of concrete and steel wherein the steel reinforcement provides the tensile strength lacking in the concrete.

* Steel reinforcing is also capable of resisting compression forces and is used in columns as well as in other situations to be described later.

1.2 <u>Advantages</u> of Reinforced Concrete as a Structural Material:

* Reinforced Concrete may be the most important material available for construction.

* It is used in one form or another for almost all structures, great or small.
<u>Examples</u>: Buildings, bridges, pavements, dams, retaining walls, tunnels, viaducts, drainage and irrigation facilities, tanks, etc.

* The tremendous success of this universal construction material can be understood quite easily if its numerous <u>advantages</u> are considered.

The advantages of reinforced concrete include the following:

1. It has considerable compressive strength as compared to most other materials.

2. Reinforced concrete has great resistance to the actions of fire and water. In fact, it is the best structural material available for situations where water is present. During fires of average intensity, members with a satisfactory cover of concrete over the reinforcing bars suffer only surface damage without failure.

3. Reinforced concrete structures are very rigid.

4. It is a low-maintenance material.

5. As compared with other materials, it has a very long service life. In fact, the strength of concrete does not decrease with time but actually increases over a very long period, measured in years, due to the lengthy process of the solidification of the cement paste.

6. It is usually the only economical material available for footings, basement walls, piers, and similar applications.

7. A special feature of concrete is its ability to be cast into an extraordinary variety of shapes from simple slabs, beams and columns to great arches and shells.

8. In most areas, concrete takes advantage of inexpensive local materials (sand, gravel, and water) and requires relatively small amounts of cement and reinforcing steel.

9. A lower grade of skilled labor is required for erection as compared to other materials such as structural steel.

1.3 Disadvantages of Reinforced Concrete as a Structural Material:

* To use concrete successfully the designer must be completely familiar with its weak points as well as with its strong ones.

* The disadvantages of reinforced concrete include the following: [3

1. Concrete has a very _low tensile strength_ , requiring the use of tensile reinforcing.

2. _Forms_ are required to hold the concrete in place until it hardens sufficiently. _Formwork_ is very _expensive_.

3. The low strength per unit weight of concrete leads to _heavy members_. This becomes an increasingly important matter for long-span structures.

4. Similarly, the low strength per unit of volume of concrete means members will be relatively _large_, an important considerations for tall buildings and long-span structures.

5. The properties of concrete vary widely due to _variations_ in its _proportioning_ and _mixing_. Furthermore, the placing and curing of concrete is _not_ as carefully _controlled_ as is the production of other materials such as structural steel and laminated wood.

6. Two others characteristics that can cause problems are concrete's _shrinkage_ and _creep_. These are discussed later.

7. Time element.

1.4 Historical Background:

* In 1824, an English bricklayer named Joseph Aspdin, after long and laborious experiments, obtained a patent for a cement which he called "portland cement".

* In the 1870s, the first portland cement was manufactured in the United States.

* In 1867, Joseph _Monier_ invented reinforced concrete.

* From 1867 to 1881, Monier received patents for reinforced concrete railroad ties, floor slabs, arches, footbridges, buildings, and other items in both France and Germany.

* In 1861, another Frenchman, François Coignet, published a book on reinforced concrete design.

* In 1875, William Ward built the first reinforced L4 concrete building in the United States, in New York.
* In 1877, Thaddeus Hyatt, an American, was probably the first person to correctly analyze the stresses in a reinforced concrete beam. He published a 28-page book on the subject.
* In the 1990s, the development and use of reinforced concrete in the world has been very rapid.

1.5 Comparison of Reinforced Concrete and Structural Steel for Buildings and Bridges.

---

* The selection of the structural material to be used for a particular building depends on the following:
  1. The height and span of the structure.
  2. the material market.
  3. the foundation conditions.
  4. the local building codes.
  5. the architectural considerations.

* Structural steel is more competitive for buildings more than 20 storeys high. Today, however, reinforced concrete is becoming increasingly competitive above 20 storeys.
* The tallest reinforced concrete building in the world is the 71-storey, 969-ft building at 311 South Wacker Drive in Chicago (U.S.A.) (295 m tall).
* If foundation conditions are poor, a lighter structural steel frame may be desirable.
* The building code in a particular city may be favorable to one material over the other. For instance, many cities have fire zones in which only fireproof structures can be erected — a very favorable situation for reinforced concrete.
* The time element favors structural steel frames, as they can be erected much more quickly than reinforced concrete ones.

1.6 Compatibility of Concrete and Steel:

* Reinforced concrete is a <u>composite</u> material made of both concrete and steel.

* Concrete and steel <u>work together</u> beautifully in reinforced concrete structures.

* The advantages of each material compensate for the disadvantages of the other.
  <u>Example</u>: The great shortcoming of concrete is its lack of tensile strength; but the tensile strength is one of the great advantages of steel.

* Reinforcing bars have tensile strengths equal to approximately 100 times that of the usual concretes used.

* The two materials <u>bond</u> together very well so there is <u>no</u> <u>slippage</u> between the two, and thus they will act together as a unit in resisting forces.

* The excellent <u>bond</u> obtained is due to the following:
  1. the chemical adhesion between the two materials.
  2. the natural roughness of the bars.
  3. the closely spaced rib-shaped deformations rolled on the bar surfaces.

* Reinforcing bars are subject to <u>corrosion</u>, but the concrete surrounding them provides them with excellent protection.

* The strength of <u>exposed</u> steel subject to the <u>temperatures</u> reached in fires of ordinary intensity is nil, but the enclosure of the reinforcement in concrete provides very satisfactory fire ratings.

* Concrete and steel work together very well in relation to temperature changes because their coefficients of thermal expansion are quite close to each other.
  For steel, the coefficient is $\alpha = 0.0000065$.
  For concrete, the coefficient varies from about $\alpha = 0.000004$ to 0.000007 (average value $= 0.0000055$).

1.7 Properties of Reinforced Concrete:

* A thorough knowledge of the properties of concrete is necessary for the student before he or she begins to design reinforced concrete structures.

Compressive Strength: f_c' :

* The <u>compressive strength</u> of concrete (f_c') is determined by testing to failure 28-day-old concrete <u>cylinders</u> (or <u>cubes</u>) at a specified rate of loading.

* Although concretes are available with 28-day ultimate strengths from 2500 psi (17.2 KN/m^2) up to as high as $10,000$ to $20,000$ psi (70 to 140 KN/m^2), most of the concretes used fall into the 3000-to-7000 psi (20-to-50 KN/m^2) range.

* For some applications, such as that for the columns of the lower storeys of high-rise buildings, concretes with strengths up to 9000 psi (60 KN/m^2) have been used.

* The <u>field conditions</u> are not the same as those in the curing room (in the laboratory), and the 28-day strengths described above cannot be achieved in the field unless almost perfect proportioning, mixture, vibration, and moisture conditions are present.

* As a result, Section 5.3 of the ACI Code requires that the concrete compressive strengths used as a basis for selecting the concrete proportions must exceed the specified 28-day strengths by anywhere from a few hundred psi (between 0 – 10 KN/m^2) up to as much as 1400 psi (10 KN/m^2), the actual amount depending on the quality control records of the concrete plant.

* Studying the stress-strain curve of reinforced concrete reveals the following:

1. The curves are roughly straight while the load is increased from zero to about one-third to one-half of the concrete's ultimate strength.

2. Beyond this range, the behavior of concrete is nonlinear.

3. Regardless of the strength, concrete reaches its ultimate strength at strains of about $\varepsilon = 0.002$

4. Concrete does not have a definite yield point (strength); rather, the curves run smoothly on to the point of rupture at strains from 0.003 to 0.004

* It will be assumed that concrete fails at $\varepsilon = 0.003$.

5. Many tests have clearly shown that stress-strain curves of concrete (cylinders or cubes) are almost identical with those for the compression sides of beams.

6. It is noticed that the weaker grades of concrete are less brittle than the stronger ones — that is, they will take larger strains before breaking.

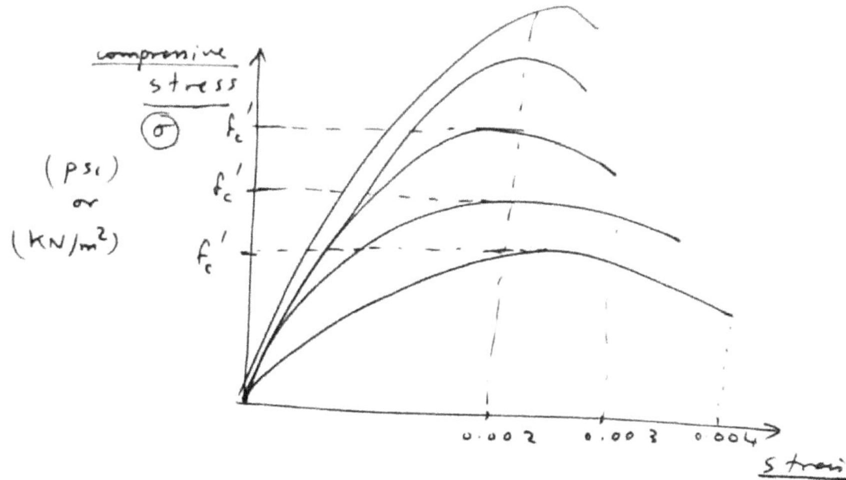

Modulus of Elasticity: E :

* Concrete has no clear-cut modulus of elasticity. Its value varies with different concrete strengths, concrete age, type of loading, and the characteristics of the cement and aggregates.

* Furthermore, there are several different definitions of the modulus:

1. The initial modulus is the slope of the stress-strain curve at the origin of the curve.

the 200-mm cube strength [1.31]. For lightweight concrete, cylinder strength and cube strength are nearly equal.

In the United States, when taking test cores of concrete in existing structures, specimens less than the standard 6 in. diameter but having a 2:1 height-to-diameter ratio are usually cut from the concrete.

An interesting discussion of the cylinder test is given by Shilstone [1.32]. ACI Committee 214 has a recommended practice [1.34] for evaluating strength test results, and Tait [1.33] has discussed the use of test results. When an assessment of the strength of in-place concrete is desired, procedures ranging from tests of cylindrical cores cut from the structure to the use of nondestructive tests [1.35–1.41] are available.

The stress–strain behavior of concrete is dependent on its strength, age at loading, rate of loading, aggregates and cement properties, and type and size of specimens [1.42, 1.43]. Typical curves for specimens loaded in compression at 28 days using normal testing speeds are shown in Fig. 1.7.2.

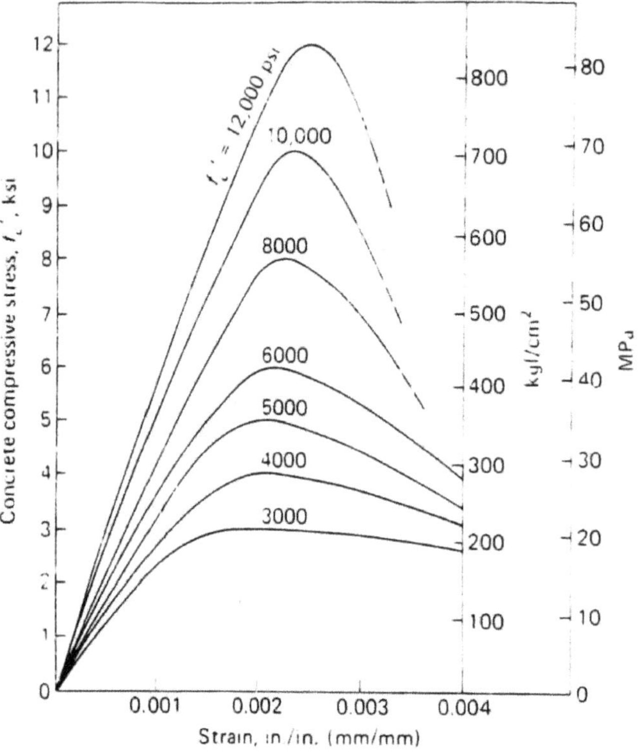

Figure 1.7.2 Typical stress–strain curves for concrete in compression under short-time loading. (Curves represent a compromise adapted from curves and results given by Wang, Shah, and Naaman [1.43], Bertero [1.44], Naaman [1.45], and Nilson [1.46].)

2. The <u>tangent modulus</u> is the slope of a tangent to the curve at some point along the curve — for instance, at 50% of the ultimate strength of the concrete.

3. The <u>secant modulus</u> is the slope of a line drawn from the origin to a point on the curve somewhere between 25% and 50% of its ultimate compressive strength.

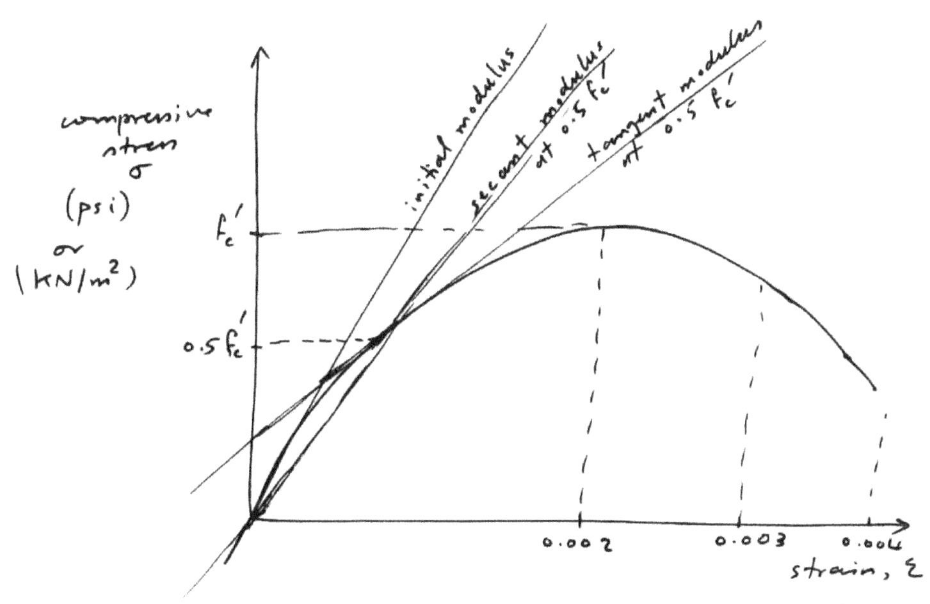

(Stress-Strain Curve for Concrete in Compression).

---

* Section 8.5 of the ACI Code states that the following expression can be used for calculating the modulus of elasticity of concretes weighing from $w_c = 90$ to $w_c = 155$ lb/ft³ :

$$E_c = 33 \, w_c^{1.5} \sqrt{f_c'}$$

where w_c is the weight of concrete in lb/ft³ (pcf), and f_c' and E_c are in psi.

* In SI units, for $w_c = 1500$ kg/m³ to $w_c = 2500$ kg/m³, the formula is:

$$E_c = 0.043 \, w_c^{1.5} \sqrt{f_c'} \qquad (ACI - 8.5.1)$$

where w_c is the weight of concrete in kg/m³, and f_c' and E_c are in MPa ($\cong 1000$ kN/m²).

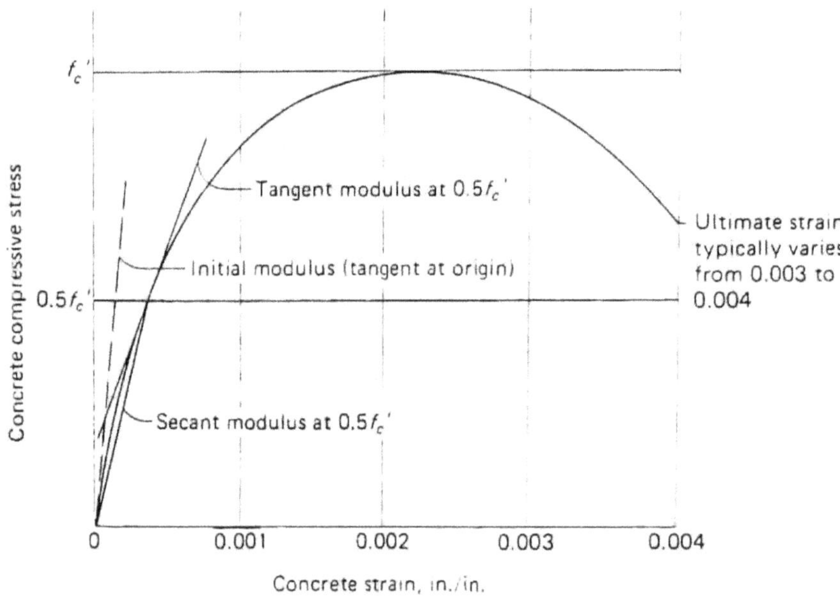

Figure 1.9.1 Stress–strain curve for concrete in compression.

$$E_c = 33w_c^{1.5}\sqrt{f_c'} \qquad (1.9.1)^*$$

was developed [1.56] for values of w_c between 90 and 155 lb/cu ft. Equation (1.9.1) may be considered as the secant modulus for a compressive stress at service load level. (A recent review of the applicability of Eq. (1.9.1) has been made by Shih, Lee, and Chang [1.82].) For normal-weight concrete weighing 145 pcf, the formula gives $E_c = 57{,}600\sqrt{f_c'}$. For normal-weight concrete ACI-8.5.1 suggests

$$E_c = 57{,}000\sqrt{f_c'} \qquad (1.9.2)^\dagger$$

Values of modulus of elasticity for various concrete strengths appear in Table 1.9.1.

— — — —

* For SI, with w_c in kg/m³ and E_c and f_c' in MPa,

$$E_c = 0.043w_c^{1.5}\sqrt{f_c'} \quad \text{(ACI 318–89M)} \qquad (1.9.1)$$

† For SI, with E_c and f_c' in MPa,

$$E_c = 5000\sqrt{f_c'} \quad \text{(ACI 318–89M)} \qquad (1.9.2)$$

and with E_c and f_c' in kgf/cm²,

$$E_c = 15{,}000\sqrt{f_c'} \quad \text{(approximate)} \qquad (1.9.2)$$

* The above formula gives the _secant modulus_ with the line drawn from the origin to a point on the stress-strain curve corresponding approximately to the stress $0.45 f_c'$.

* For normal weight concrete weighing approximately $145 \ lb/ft^3$ $(22.8 \ kN/m^3)$, the ACI Code suggests the following simplified version of the previous expression:

$$E_c = 57,000 \sqrt{f_c'}$$

where f_c' and E_c are in psi.

In SI units, we have the equivalent formula:

$$E_c = 4700 \sqrt{f_c'}$$

$$\underline{\underline{or}} \ \text{approximately} \quad E_c = 5000 \sqrt{f_c'} \qquad\qquad (ACI-8.5.1)$$

where f_c' and E_c are in MPa.

$$\left[\underline{\underline{or}} \quad E_c = 15,000 \sqrt{f_c'}, \quad \text{when } f_c' \text{ and } E_c \text{ are in } kg/cm^2. \right]$$

* The following table shows values of E_c for different-strength concretes (calculated at $w_c = 145 \ lb/ft^3$ or $w_c = 22.8 \ kN/m^3$).

English Units		SI Units	
f_c' (psi)	E_c (psi)	f_c' (MPa)	E_c (MPa)
3000	3,150,000	21	22,900
3500	3,400,000	24	24,500
4000	3,640,000	28	26,500
4500	3,860,000	31	27,800
5000	4,070,000	35	29,600

calculated using
$$E_c = 57,000 \sqrt{f_c'}$$

calculated using
$$E_c = 5000 \sqrt{f_c'}.$$

Table 1.9.1 Values of Modulus of Elasticity (Using $E_c = 33w_c^{1.5}\sqrt{f_c'}$ for Normal-Weight Concrete Weighing 145 pcf)

INCH-POUND UNITS		SI UNITS[b]	
f_c' (psi)	E_c (psi)	f_c' (MPa)	E_c' (MPa)
3000	3,150,000	21[a]	22,900
3500	3,400,000	24	24,500
4000	3,640,000	28	26,500
4500	3,860,000	31	27,800
5000	4,070,000	35	29,600

[a] These metric values are rounded values approximating concrete strengths in Inch-Pound units. actual equivalents for 3000. 3500. 4000. 4500. and 5000 psi are 20.7, 24.1, 27.6, 31.0, and 34.5 MPa. respectively

[b] Multiply MPa values by 10.2 to obtain kgf/cm².

[c] Using $E_c = 5000\sqrt{f_c'}$ as per ACI 318-83M

1.10 CREEP AND SHRINKAGE

Creep and shrinkage are time-dependent deformations. that. along with cracking provide the greatest concern for the designer because of the inaccuracies and unknowns that surround them. Concrete is elastic only under loads of short duration and. because of additional deformation with time. the effective behavior is that of an inelastic material. Deflection after a long period of time is therefore difficult to predict. but its control is needed to assure serviceability during the life of the structure

Creep

Creep is the property of concrete (and other material) by which it continues to deform with time under sustained loads at unit stresses within the accepted elastic range (say. below $0.5f_c'$). This inelastic deformation increases at a decreasing rate during the time of loading, and its total magnitude may be several times as large as the short-time elastic deformation. Frequently creep is associated with shrinkage. since both are occurring simultaneously and often provide the same net effect: increased deformation with time. As may be noted by the general relationship of deformation versus time in Fig. 1.10.1. the "true elastic strain" decreases since the modulus of elasticity E_c is a function of concrete strength f_c' which increases with time

Although creep is separate from shrinkage. it is related to it. Detailed information is available for predicting creep [1.57, 1.58]. The internal mechanism of creep. or "plastic flow" as it is sometimes called. may be due to any one or a combination of the following: (1) crystalline flow in the aggregate and hardened cement paste.

Poisson's Ratio: ν :

* As a concrete cylinder (or cube) is subjected to compressive loads, it not only shortens in length, but also expands laterally. Poisson's ratio (ν) is the ratio of the lateral expansion to the longitudinal shortening.

* Its value ranges from about $\nu = 0.11$ for the higher-strength concrete to as high as $\nu = 0.21$ for the weaker-grade concretes, with average values of $\nu = 0.16$.

Tensile Strength:

* The tensile strength of concrete varies from about 8% to 15% of its compressive strength.

* A major reason for this small strength is the fact that concrete is filled with fine cracks. The cracks have little effect when concrete is subjected to compression loads because the loads cause the cracks to close and permit compression transfer. You can see this is not the case for tensile loads.

* The tensile strength of concrete is normally neglected in design calculations.

* Concrete is not assumed to resist a portion of the tension in a flexural member and the steel the remainder. The reason is that concrete cracks at such small tensile strains that the low stresses in the steel up to that time would make its use uneconomical.

* The concrete tensile strength is important in:
  1. affecting the sizes and extent of the cracks that occur.
  2. reducing the deflections.

* The tensile strength of concrete doesn't vary in direct proportion to its ultimate strength f_c'. It does, however, vary approximately in proportion to $\sqrt{f_c'}$.

* The tensile strength of concrete is quite difficult to measure with direct axial tension loads because of problems in gripping test specimens so as to avoid stress concentrations and because of difficulties in aligning the loads.

* As a result of these problems, two rather indirect tests[11] have been developed to measure concrete's tensile strength.

1. **the modulus of rupture test:**

* Based on hundreds of tests, the ACI Code - (9.5.2.3) provides a modulus of rupture f_r equal to:

$$f_r = 7.5 \sqrt{f_c'} \quad, \quad \text{where } f_c', f_r \text{ are in } \underline{psi}.$$

* In SI units, the modulus of rupture, f_r, is given by:

$$f_r = 0.62 \sqrt{f_c'} \quad, \quad \text{where } f_c', f_r \text{ are in } \underline{MPa}.$$

* <u>Note</u>: the modulus of rupture is the tensile strength <u>in</u> <u>flexure</u>, calculated using the flexure formula:

$$f_r = \frac{Mc}{I}$$

2. **The split-cylinder strength test:**

* The split-cylinder tensile strength f_{ct} has been found to be proportional to $\sqrt{f_c'}$, such that:

$$f_{ct} = 6\sqrt{f_c'} \text{ to } 7\sqrt{f_c'} \quad \text{for normal-weight concrete}$$

$$f_{ct} = 5\sqrt{f_c'} \text{ to } 6\sqrt{f_c'} \quad \text{for lightweight concrete.}$$

Both f_{ct} and f_c' are in psi.

* In SI Units, we have:

$$f_{ct} = 0.5\sqrt{f_c'} \text{ to } 0.6\sqrt{f_c'} \quad \text{for normal-weight concrete}$$

$$f_{ct} = 0.4\sqrt{f_c'} \text{ to } 0.5\sqrt{f_c'} \quad \text{for lightweight concrete.}$$

* <u>Note</u>: the split-cylinder strength is the tensile strength at which splitting occurs in the test cylinder.

<u>Shear Strength:</u>

* It is extremely difficult in testing to obtain pure shear failures unaffected by other stresses. As a result, the tests of concrete shearing strengths have yielded values all the way from one-third to four-fifths of the ultimate compressive strengths.

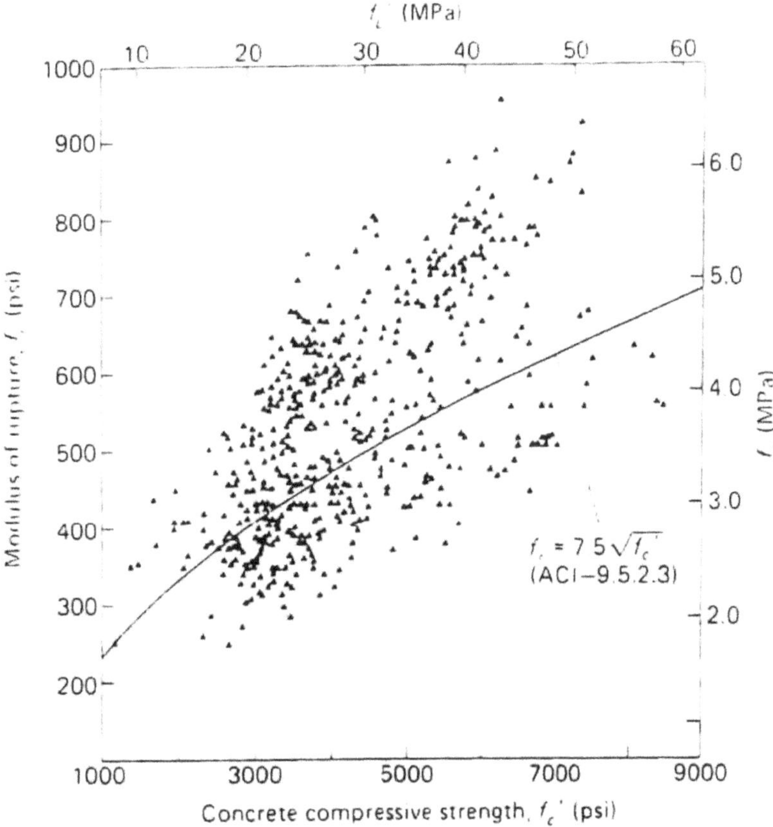

Figure 1.8.1 Comparison of test results for modulus of rupture with ACI Code expression (adapted from Mirza, Hatzinikolas, and MacGregor [1.55])

1.9 MODULUS OF ELASTICITY

The modulus of elasticity of concrete varies, unlike that of steel, with strength depends, though to a much lesser extent, on the age of concrete, properties aggregates and cement, rate of loading, and the type and size of specimen. F more, since concrete exhibits some permanent set even under small loads, th various definitions of the modulus of elasticity.

Referring to Fig. 1.9.1, representing a typical stress–strain curve for con compression, the initial modulus, the tangent modulus, and the secant modu noted. Usually the secant modulus at from 25 to 50% of the compressive s f_c' is considered to be the modulus of elasticity. For many years the modu approximated adequately as $1000f_c'$ by the ACI Code; but with the rapidly inc use of lightweight concrete, the variable of density needed to be included. As of a statistical analysis of available data, the empirical formula given by AC

Table 1.12.2   ASTM (and Canadian Standard) Metric Reinforcing Bar Dimensions and Weights in SI Units

BAR NUMBER	DIAMETER (mm)	AREA (cm^2)	WEIGHT (kg/m)	COMPARISON TO U.S. BARS BAR NUMBER (AREA)
10M	11.3	1.0	0.784	— 3 (0.71 cm^2)
15M	16.0	2.0	1.568	— 4 (1.29 cm^2)
20M	19.5	3.0	2.352	— 5 (2.00 cm^2)
25M	25.2	5.0	3.920	— 6 (2.84 cm^2)
30M	29.9	7.0	5.488	— 7 (3.87 cm^2)
35M	35.7	10.0	7.840	— 8 (5.10 cm^2)
45M	43.7	15.0	11.760	— 9 (6.45 cm^2)
55M	56.4	25.0	19.600	—10 (8.19 cm^2)
				—11 (10.06 cm^2)
				—14 (14.52 cm^2)
				—18 (25.81 cm^2)

Table 1.12.3   Reinforcing Bar Steels

ASTM DESIGNATION	GRADE	BAR SIZES	MINIMUM YIELD STRESS[a] f_y		MINIMUM TENSILE STRENGTH	
			ksi	MPa	ksi	MPa
A615	40	#3–#6	40		70	
(Billet steel)	60	#3–#18	60		90	
	75	#11, #14, #18	75		100	
A615[b]	300	10–20		300		500
(Billet steel	400	10–55		400		600
	500	35, 45, 55		500		700
A616/A616M	50	#3–#11	50		80	
(Rail steel)	60	#3–#11	60		90	
	350	10–35		350		550
	400	10–35		400		600
A617/A617M[c]	40	#3–#11	40		70	
(Axle steel)	60	#3–#11	60		90	
	300	10–35		300		500
	400	10–35		400		600
A706/A706M	60	#3–#18	60		80	
(Low-alloy steel)	400	10–55		400		550

[a] The term "yield stress" refers to either *yield point*, the well-defined deviation from perfect elasticity, or *yield strength*, the value obtained by a specified offset strain for material having no well-defined yield point.

[b] Metric (SI) specification applies to bars designated numbers 10 through 55, as given in Table 1.12.2

[c] Although rail steel is no longer considered a practical source of bar reinforcement, its use is still permitted provided the bend test requirements for axle steel, ASTM 617/617M, Grade 60, can be satisfied.

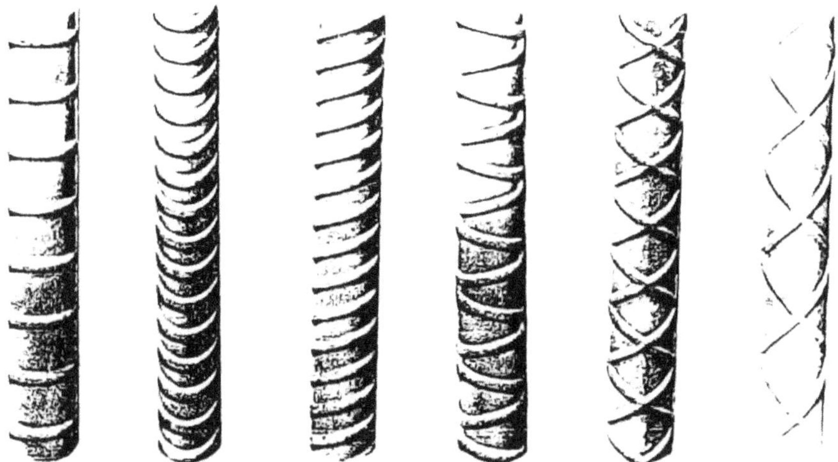

Figure 1.12.1 Deformed reinforcing bars. (Courtesy of Concrete Reinforcing Steel Institute.)

inch included in the nominal diameter of the bars. Bars of designation #9 through #11 are round bars corresponding to the former 1 in. square, $1\frac{1}{8}$ in. square, and $1\frac{1}{4}$ in. square sizes, and bars designated #14 and #18 are round bars having cross-sectional areas equal to those of $1\frac{1}{2}$ and 2 in. square sizes, respectively. The nominal

Table 1.12.1 Standard Reinforcing Bar Dimensions and Weights (Bars in inch-Pound Units According to ASTM A615 [1.65], A616 [1.66], A617 [1.67], and A706 [1.68])

	NOMINAL DIMENSIONS				WEIGHT	
BAR	DIAMETER		AREA			
NUMBER	(in.)	(mm)	(sq in.)	(cm²)	(lb/ft)	(kg/m)
3	0.375	9.5	0.11	0.71	0.376	0.559
4	0.500	12.7	0.20	1.29	0.668	0.994
5	0.625	15.9	0.31	2.00	1.043	1.552
6	0.750	19.1	0.44	2.84	1.502	2.235
7	0.875	22.2	0.60	3.87	2.044	3.041
8	1.000	25.4	0.79	5.10	2.670	3.973
9	1.128	28.7	1.00	6.45	3.400	5.059
10	1.270	32.3	1.27	8.19	4.303	6.403
11	1.410	35.8	1.56	10.06	5.313	7.906
14	1.693	43.0	2.25	14.52	7.65	11.38
18	2.257	57.3	4.00	25.81	13.60	20.24

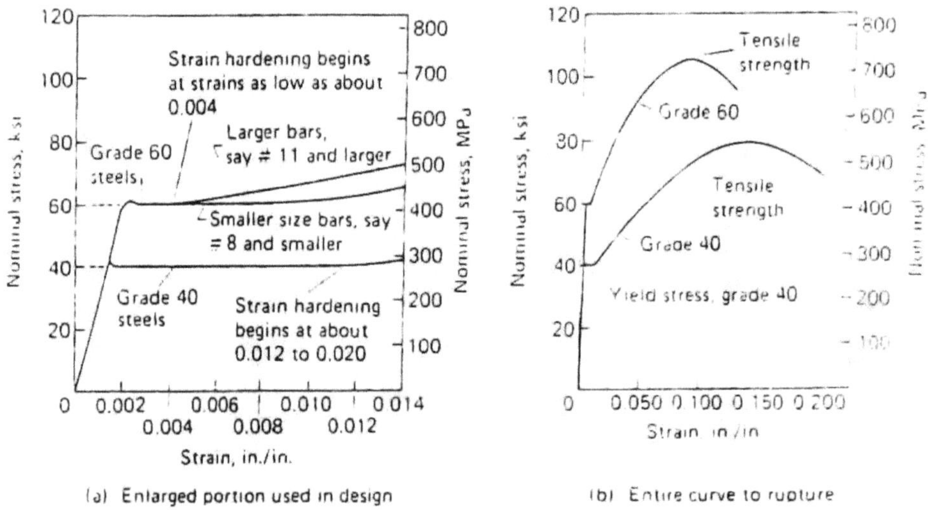

Figure 1.12.2 Typical stress-strain curves for reinforcing bar steels in tension

diameter of a deformed bar is equivalent to the diameter of a plain bar having the same weight per-foot as the deformed bar.

The ASTM standard sizes for metric reinforcing bars agree with the Canadian standard sizes for such bars, the designations and dimensions of which are given in Table 1.12.2.

Reinforcing bar steel in the United States is covered under ASTM designations as shown in Table 1.12.3. The "Grade" of steel is the minimum specified yield stress* expressed in ksi for Inch-Pound reinforcing bar Grades 40, 50, 60, and 75, and in MPa for SI reinforcing bar Grades 300, 350, 400, and 500.

Billet steel (ASTM A615 and A615M) is newly made steel having its chemical content sufficiently controlled to provide necessary ductility. Axle and rail steel bars, both of which are rarely used now, are rerolled from old axles and rails, and are generally less ductile than bars of billet steel.

Grade 60 steel is the primary reinforcement material. Formerly, Grade 40 was more economical and readily available; today (1992), Grade 60 is economical and Grade 40 billet steel is available only in the smaller bars #3 through #6. Both Grades 40 and 60 exhibit the well-defined yield point and elastic–plastic strain behavior shown in Fig. 1.12.2(a). The overall relationships are shown in Fig. 1.12.2(b).

Welded wire fabric is used in thin slabs, thin shells, and other locations where available space would not permit the placement of deformed bars with proper cover and clearance. Welded wire fabric (WWF) covered under ASTM A185 [1.70] and

---

* The term "yield stress" refers to either *yield point,* the well-defined deviation from perfect elasticity, or *yield strength,* the value obtained by a specified offset strain for material having no well-defined yield point.

Table 9: Weight, Perimeter and Area of Metric Steel Bars

Bar diameter,	Weight per meter,	Peri-meter,	Area in cm² for various numbers of bars											
mm	kg	cm	1	2	3	4	5	6	7	8	9	10	11	12
5	0.15	1.57	0.20	0.39	0.59	0.78	0.98	1.18	1.37	1.57	1.77	1.96	2.16	2.35
6	0.22	1.88	0.28	0.56	0.85	1.13	1.41	1.70	1.98	2.26	2.54	2.83	3.11	3.40
8	0.39	2.51	0.50	1.00	1.51	2.01	2.51	3.01	3.52	4.02	4.52	5.03	5.53	6.04
10	0.62	3.14	0.79	1.57	2.36	3.14	3.93	4.71	5.50	6.28	7.07	7.85	8.64	9.42
12	0.89	3.77	1.13	2.26	3.39	4.52	5.65	6.78	7.91	9.05	10.18	11.31	12.44	13.57
14	1.21	4.40	1.54	3.08	4.62	6.16	7.70	7.24	10.78	12.32	13.86	15.39	16.94	18.48
16	1.58	5.03	2.01	4.02	6.03	8.04	10.05	12.06	14.07	16.08	18.09	20.11	22.12	24.13
18	2.00	5.65	2.54	5.09	7.63	10.18	12.72	15.27	17.81	20.36	22.90	25.45	27.99	30.54
20	2.47	6.28	3.14	6.28	9.42	12.57	15.71	18.84	21.99	25.14	28.28	31.42	34.56	37.70
22	2.98	6.91	3.80	7.60	11.40	15.21	19.01	22.81	26.61	30.41	34.21	38.01	41.81	45.62
25	3.85	7.85	4.91	9.82	14.73	19.63	24.54	29.45	34.36	39.27	44.18	49.09	54.00	58.91
28	4.83	8.80	6.16	12.31	18.47	24.63	30.79	36.94	43.10	49.26	55.42	61.58	67.73	73.89
30	5.54	9.42	7.07	14.14	21.20	28.27	35.34	42.41	49.48	56.54	63.61	70.68	77.75	84.82
32	6.31	10.05	8.04	16.08	24.13	32.17	40.21	48.26	56.30	64.34	72.38	80.42	88.47	96.51
36	7.99	11.31	10.18	20.36	30.54	40.72	50.90	61.07	71.26	81.43	91.61	101.79	111.97	122.15
40	9.86	12.57	12.57	25.13	37.70	50.26	62.83	75.40	87.96	100.53	113.09	125.66	138.23	150.80
45	12.49	14.14	15.90	31.81	47.71	63.62	79.52	95.42	111.33	127.32	143.14	159.04	174.94	190.85
50	15.41	15.71	19.64	39.27	58.91	78.54	98.18	117.81	137.45	157.08	176.72	196.35	215.99	235.62

1.8 Steel Reinforcement:

* Steel reinforcement may consist of:
  1. bars.
  2. welded wire fabric.
  3. wires.
* For usual construction, bars (called <u>deformed bars</u>) having lugs and protrusions (deformations) are used.
* Such deformations inhibit the longitudinal movement of the bar relative to the concrete that surrounds it.
* These deformed bars are available in different sizes.
* Grade 40 and Grade 60 exhibit the well-defined yield point and elastic-plastic strain behavior shown in the figure.

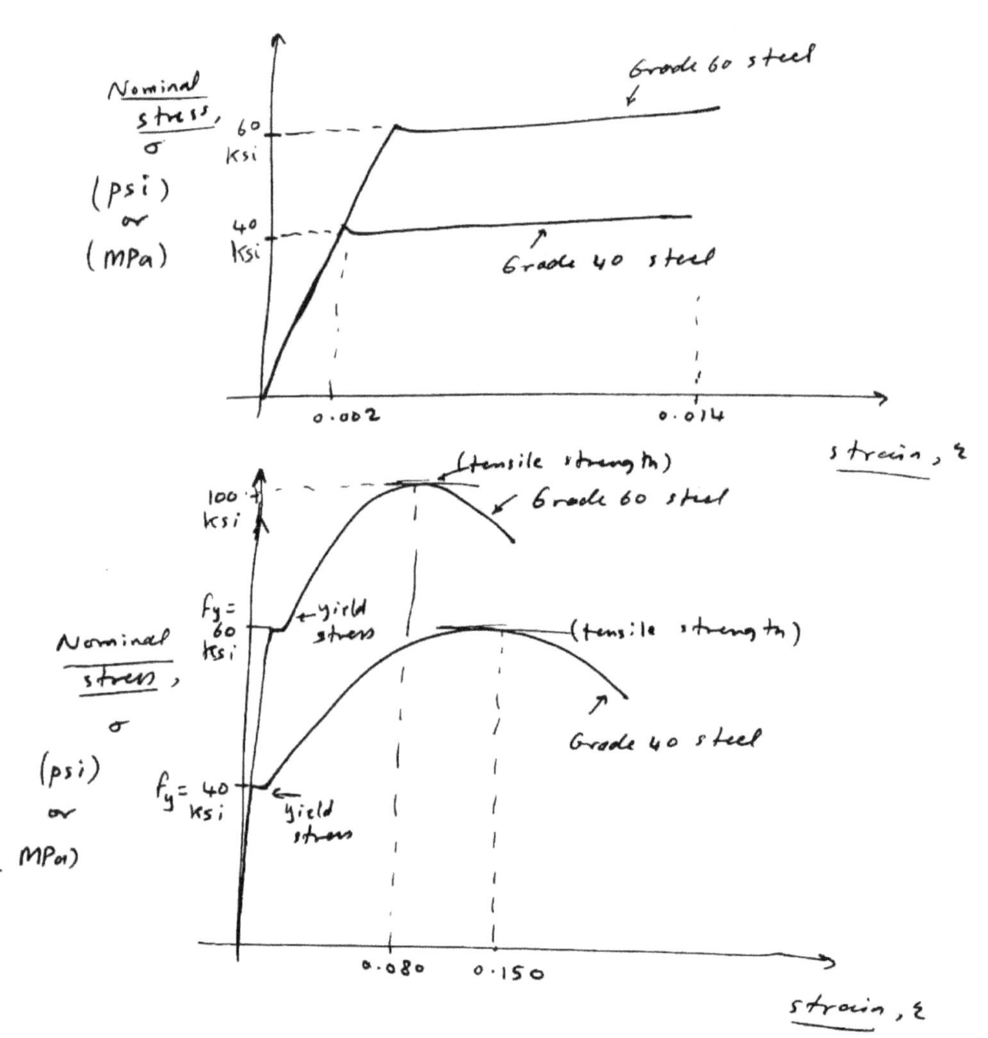

* The modulus of elasticity E_s for steel may be taken as 29,000,000 psi (200,000 MPa).

Chapter 2
Design Methods

2.1 The ACI Building Code:

* With two materials, such as steel and concrete, act together, it is understandable that the analysis for strength of a reinforced concrete member has to be partly empirical, although mostly rational.

* These semi-rational principles and methods are being constantly revised and improved as a result of theoretical and experimental research accumulate.

* The American Concrete Institute (ACI) issues building codes requirements such as their two latest ones:

  1. Building Code Requirements for Reinforced Concrete, (ACI-318-89), revised in 1992.

  2. Building Code Requirements for Structural Concrete, (ACI-318-95), 1995.

* These codes will be referred to as the ACI Code.

* The student is advised to have the ACI Code as a ready reference in this course.

* The ACI Code is a standard of the American Concrete Institute.

* The ACI Code is partly a:

  1. specification-type code: which gives acceptable design and construction methods in detail.

  2. performance code: which states desired results rather than the details of how such results are to be obtained.

* A building code, legally adopted, is intended to prevent people from being harmed; therefore, it specifies minimum requirements consistent with good safety.

* Other codes available are:
  1. British Standard BS 8110.
  2. Arab Code.
  3. Egyptian Code.
  4. Indian Code.
  5. European Code.
  6. Jordanian Code

2.2 Design Methods:

* Two philosophies of design have long been prevalent:

1. The Working Stress Method: This method focuses on the conditions at service loads (that is, when the structure is being used). This method was the principal method of design used from the early 1900s until the early 1960s.

2. The Strength Design Method: this method focuses on the conditions at loads greater than service loads when failure may be imminent. Today, in the 1990s, the strength design method is used. It is deemed conceptually more realistic to establish structural safety.

Working Stress Method:

* In the working stress method (referred to by the present ACI Code as the "alternate design method"), a structural element is so designed that the stresses resulting from the action of service loads (also called working loads) and computed by the mechanics of elastic members do not exceed some predesignated allowable values.

* The working stress method may be expressed by the following:

$$\boxed{f \leq [\text{allowable stress}, f_{\text{allow.}}]}$$

where $f \equiv$ an elastically computed stress, such as by using the flexure formula $f = \dfrac{Mc}{I}$ for a beam.

$f_{\text{allow.}} \equiv$ a limiting stress prescribed by a building code as a percentage of the compressive strength f_c' for concrete, or of the yield stress f_y for the steel reinforcing bars.

* Some of the obstacles to the working stress method are:

1. Since the limitation is on the total stress under service load, there is no simple way to account for different degrees of uncertainty of various kinds of loads.

2. Creep and shrinkage are <u>not</u> easily accounted for by the elastic calculation of stresses.

3. Concrete stress is <u>not</u> proportional to strain up to its crushing strength, so that the inherent safety provided is unknown when a percentage of f_c' is used as the allowable stress.

Strength Design Method :

* In the strength design method (formerly called ultimate strength method), the service loads are increased sufficiently by factors to obtain the load at which failure is considered to be "imminent". This load is called the <u>factored load</u> or <u>factored service load</u>.

* The structure or structural element is then proportioned such that the strength is reached when the factored loads are acting. The computation of this strength takes into account the nonlinear stress-strain behavior of concrete.

* The strength design method may be expressed by the following :

$$\boxed{\text{strength provided} \geq \begin{bmatrix} \text{strength required to} \\ \text{carry factored loads} \end{bmatrix}}$$

where the "strength provided" (such as moment strength) is computed in accordance with rules and assumptions of behavior prescribed by a building code, and the "strength required" is that obtained by performing a structural analysis using factored loads.

2.3 Advantages of Strength Design :

* Among the several advantages of the strength design method as compared to the alternate design method (working stress method) are the following :

1. The derivation of the strength design expressions takes into account the nonlinear shape of the stress-strain curve.

2. With strength design, a more consistent theory is used throughout the designs of reinforced concrete structures.

3. A more realistic factor of safety is used in strength design. In particular, the use of different load or safety factors in strength design for the different types of loads is a definite improvement.

4. A structure designed by the strength method will have a more uniform safety factor against collapse throughout. The strength method takes considerable advantage of higher-strength steels, whereas working-stress design only partly does so. The result is better economy for strength design.

5. The strength method permits more flexible designs than does the alternate design method. As a result, large sections may be used with small percentages of steel or small sections may be used with large percentages of steel.

2.4 Comments on Design Methods:

* The ACI Code uses the strength design method. It relegates the working stress method to a small section in the Code (Appendix A) and calls it the alternate design method

* Serviceability must also be considered. Serviceability factors that may be as important as strength are excessive deflection, detrimental cracking, excessive amplitude or undesirable frequency of vibration, and excessive noise transmission.

* The designer must consider both strength and serviceability

2.5 Design Handbooks and Computer Software:

* Design handbooks are available that include curves and tables that greatly speed up design.

* Computer programs are used these days for design.

2.6 Types of Loads (Service Loads):

* The most important and most difficult task faced by the structural designer is the accurate estimation of the loads that may be applied to a structure during its life.

* No loads that may reasonably be expected to occur may be overlooked.

* After loads are estimated, the next problem is to decide the worst possible combinations of these loads that might occur at one time.

* There are several types of "service" loads that include the following:

Dead Loads : D :

* Dead loads are loads of constant magnitude that remain in one position.

* Dead loads consist of the structural frame's own weight and other loads that are permanently attached to the frame.

* Example: For a reinforced concrete building, some dead loads are the frame, walls, floors, roofs, plumbing, and fixtures.

* To design a structure, it is necessary for the weights or dead loads of the various parts to be estimated for use in the analysis. The exact sizes and weights of the parts are not known until the structural analysis is made and the members of the structure selected. The weights, as determined from the actual design, must be compared with the estimated weights. If large discrepancies are present, it will be necessary to repeat the analysis and design using better estimated weights.

* Reasonable estimates of structure weights may be obtained by referring to similar-type structures or to various formulas and tables available in most civil engineering handbooks.

* An experienced designer can estimate very closely the weights of most structures and will spend little time repeating designs because of poor estimates.

<u>Live Loads: L:</u>

* Live loads are loads that may change in position and magnitude.
* Simply stated, all loads that are not dead loads are live loads.
* A brief description of some live loads follows.

<u>1. Snow and Ice Loads</u>:

* In cold regions, snow and ice loads are often very important.
* Snow load is a variable load which may cover an entire roof or only part of it.

<u>2. Traffic Loads for Bridges</u>:

* Bridges are subjected to a series of concentrated loads of varying magnitude caused by groups of truck or train wheels.

<u>3. Impact Loads</u>:

* Impact Loads are caused by the vibration of moving or movable loads.

<u>4. Lateral Loads</u>:

* Lateral loads are of two main types:

<u>4.1 Winds Loads: W</u>:

* The magnitude of wind loads vary with geographical locations, heights above ground, types of terrain surrounding the buildings including other nearby structures, and other factors.
* Wind forces act as <u>pressures</u> on vertical windward surfaces, pressures or <u>suction</u> on sloping <u>windward</u> surfaces (depending on the slope), and suction on flat surfaces and on <u>leeward</u> vertical and sloping surfaces (due to the creation of negative pressures or vacuums).

<u>4.2 Earthquake Loads: E</u>:

* Many areas of the world fall in earthquake territory, and in these areas it is necessary to consider <u>seismic</u> forces in designs for buildings.

* During an earthquake there is an acceleration of the ground surface. This acceleration can be broken down into vertical and horizontal components. Usually the vertical component of the acceleration is assumed to be negligible, but the horizontal component can be severe.

* The effect of the horizontal acceleration increases with the distance above the ground because of the "whipping effect" of the earthquake, and design loads should be increased accordingly.

5. <u>Other Live Loads</u>:

* <u>Soil pressures</u>: the exertion of lateral earth pressures on walls or upward pressures on foundations.

* <u>Hydrostatic pressures</u>: water pressure on dams, uplift pressures on tanks and basement structures.

* <u>Blast loads</u>: caused by explosions, sonic booms, and military weapons.

* <u>Thermal forces</u>: due to changes in temperature resulting in structural deformation and resulting structural forces.

* <u>Centrifugal forces</u>: caused on curved bridges by trucks and trains or similar effects on roller coasters.

2.7 Selection of Design Loads:

* There are many publications for specifications of the magnitudes of the different types of loads. These publications are put together in the form of building codes and specifications. Some of them include the following:

1. "Minimum Design Loads for Buildings and Other Structures", American Society of Civil Engineers, ASCE-7-88, 1988, N.Y., USA.

2. "Uniform Building Code", 1994, International Conference of Building Officials. Whittier, CA, USA.

3. "Standard Specifications for Highway Bridges", 1989. American Association of State Highway and Transportation Officials, AASHTO, Washington, D.C., USA.

2.8 The ACI Design Procedure:

* In the strength design method, members are sized based on <u>factored loads</u> that are greater than the service loads.

* The factored load is produced by multiplying the service load by <u>load factors</u>, numbers greater than 1.

* The size of the load factor, which represents part of the factor of safety applied to loads, reflects the accuracy with which the design loads can be predicted.

* A load whose magnitude and distribution can be established with certainty is increased by a small load factor than a load whose magnitude is subject to variation or whose exact intensity cannot be predicted with precision.

* <u>Example</u> : Dead load, which can usually be computed very accurately, is increased by a load factor of 1.4 & Live load, which is subject to greater variation, is increased by a load factor of 1.7 .

* The factored load is denoted by the symbol U .

LOAD FACTORS:	
Type of Load	ACI Load Factor
Dead Load D	1.4
Live Load L	1.7
Wind Load W	1.7
Earthquake Load E	Substitute 1.1E for W
Earth Pressure H	1.7
Fluid Pressure F	1.4
Impact Load I	Substitute (L+I) for L
Settlement, Creep, Shrinkage, Temperature T	1.4

* Factored loads or forces produced by factored loads are subscripted by a lowercase u .
  <u>Example</u> : M_u may be the moment at a section produced by factored loads.

* See section 9.2 of the ACI Code for factored loads.

<u>Load Combinations:</u>

* Using factored loads, the designer carries out an elastic analysis of the structure. Structures are usually analyzed both for full gravity loads and for wind in combination with reduced gravity loads.

* When the force in a member is due to a combination of dead and live loads, the factored load U is given by:

$$\text{Factored Load} \qquad U = 1.4D + 1.7L$$

---

* When the force in a member is due to a combination of wind load W in addition to dead and live loads, the factored load U is given by:

$$U = 0.75(1.4D + 1.7L + 1.7W)$$

where the 0.75 reduction factor accounts for the improbability of having maximum wind load and live loads acting simultaneously on the structure.

* Sometimes a more severe situation arises with wind loads when live load is absent; this possibility must be considered. When live load is absent, the above formula becomes:

$$U = 0.75(1.4D + 1.7W)$$
$$= 1.05D + 1.275W$$

* Furthermore, for situations in which dead load is a gravity stabilizing effect in combination with wind (such as a tower or wall), the possibility of a reduced dead load must be considered rather than overload; thus we have:

$$U = 0.9D + 1.3W$$

---

* If earthquake effects E must be considered in design, then the factored load is given by:

$$U = 0.75(1.4D + 1.7L + 1.7(1.1E))$$
$$= 0.75(1.4D + 1.7L + 1.87E)$$
$$= 1.05D + 1.275L + 1.40E$$

* when live load is absent, we have:

$$U = 0.9 D + 1.3 (1.1 E)$$
$$= 0.9 D + 1.43 E$$

---

* When lateral earth pressure H is involved, it is treated as live load as follows:

$$U = 1.4 D + 1.7 L + 1.7 H$$

* When live load is absent, we have:

$$U = 0.9 D + 1.7 H$$

---

* When fluid pressure F is involved, it is treated as dead loads as follows:

$$U = 1.4 D + 1.7 L + 1.4 F$$

* When live load is absent, we have:

$$U = 0.9 D + 1.4 F$$

---

* If the live load is applied rapidly, as may be the case for parking structures, loading docks, warehouse floors, elevator shafts, etc., impact loads I should be considered.

In all the equations, substitute $(L + I)$ for L when impact must be considered.

---

* Where the structural effects T of differential settlement, creep, shrinkage, or temperature change may be significant, they are to be included with dead load:

$$U = 0.75 (1.4 D + 1.4 T + 1.7 L)$$

but not less than:

$$U = 1.4 (D + T)$$

the 0.75 factor is used to recognize the low probability of having these effects occur simultaneously with full dead and live load.

---

* Any structure or structural element <u>must be designed for</u> <u>the most severe</u> of any of the <u>load combinations</u> given.

* The forces in a member created by factored loads represent the <u>required strength</u> of the member.

* Members are sized so that their <u>design strength</u> will be equal to or greater than the <u>required strength</u>.

* The <u>design strength</u> is a reduced value of the <u>ultimate</u> or <u>nominal strength</u> of the cross-section.

* The <u>nominal strength</u> of a member is evaluated in accordance with provisions and assumptions specified by the ACI Code.

* Nominal strength is evaluated either:

1. analytically: by considering the state of stress associated with the particular mode of failure (either steel yields or concrete crushes),

2. experimentally: by studies that relate the ultimate strength to the proportions of the cross-section and the strength of the materials.

* Nominal strength is designated by the subscript n.

* In order to account for inevitable losses in member strength due to imperfect workmanship, e.g. undersized members, bars placed out of position, or voids in the concrete and understrength materials, the nominal strength of a member is <u>reduced</u> by multiplying by a <u>capacity reduction factor</u> ϕ, a number less than one, to give the <u>design strength</u>.

* The magnitude of the reduction factor ϕ is influenced by the ductility of the member, the degree of accuracy with which the member's capacity can be predicted, and the importance of the member to the overall strength of the structure.

* For example, a flexural failure of a beam involves one member and produces a local failure, but failure of a column may result in the collapse of many floors.

* The reduction factor ϕ constitutes the second part of
the factor of safety in strength design.

CAPACITY REDUCTION FACTORS, ϕ	
Nominal Strength	Reduction Factor ϕ
(Flexure) Bending (with or without axial tension)	0.9
Axial Tension	0.9
Shear and Torsion	0.85
Compression Members (Columns) with spirals	0.75
Compression Members (Columns) with ties, others	0.7
Bearing on Concrete	0.7
Bending in Plain Concrete	0.65

* Load factors and reduction factors have been selected for the ACI Code so that failure of structures will initiate in beams by yielding of the tension steel.

* Yielding of the tension steel in properly designed beams produces sagging and heavily cracked members but does not cause total collapse of the structure. Factors of safety against other less ductile modes of failure are much higher.

* In summary, the design criteria of the strength method can be stated as:

$$\boxed{\text{Required strength} \leq \text{Design strength}}$$

or $$\boxed{\text{Required strength} \leq \phi \cdot (\text{Nominal strength})}$$

Example: Applying the above criteria to a beam that is stressed only by shear V and moment M, we have at every section:

$$\boxed{\begin{aligned} V_u &\leq \phi V_n \\ M_u &\leq \phi M_n \end{aligned}}$$

where
$$\begin{cases} V_u \equiv \text{shear force produced by factored loads.} \\ M_u \equiv \text{moment produced by factored loads.} \end{cases}$$

<u>Required strength</u>

and
$$\begin{cases} V_n \equiv \text{nominal shear force} \\ M_n \equiv \text{nominal moment (flexural strength).} \end{cases}$$

<u>Nominal strength</u>

<u>Note</u> : Since the strength method of design is based on behavior at failure, it does not guarantee that behavior will also be satisfactory under service loads; therefore, the ACI Code has established additional criteria to ensure that members will also satisfy the requirements of serviceability (e.g. deflections, cracks, etc.).

<u>Example 1</u> :

Determine the axial forces for which the member CD in the figure should be designed given the following service loads :

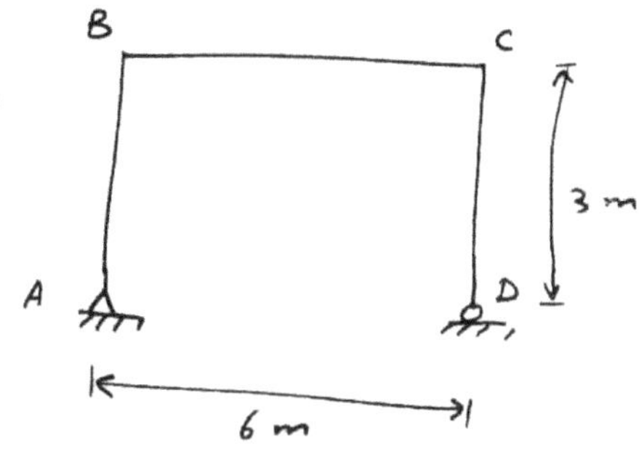

(a) dead load of 14.5 kN/m on girder BC ,

(b) live load of 30 kN/m on girder BC ,

(c) wind load of 90 kN horizontally at joint B.

Wind may act in either direction .

<u>Solution</u> :

For section 1-1,

$+\uparrow \ \Sigma F_y = 0 :$

$$R_D + N_{CD} = 0$$

$$\boxed{N_{CD} = - R_D}$$

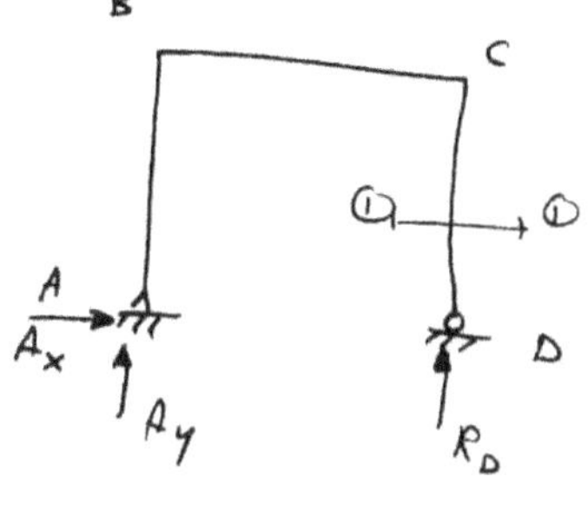

∴ the axial force in member CD is the negative of the reaction at the roller at D.

(a) $D = 14.5 \text{ kN/m}$ on BC.
(b) $L = 30 \text{ kN/m}$ on BC.
(c) $W = 90 \text{ kN}$, horizontal force at B, in any direction.

Consider the following cases of load combinations:

<u>Case 1</u>: $L+D$ only:

$$w_u = 1.4D + 1.7L = 1.4(14.5) + 1.7(30) = 71.3 \text{ kN/m}.$$

From symmetry,

$$R_D = \tfrac{1}{2}(w_u)(6) = 3 w_u.$$

$$\therefore \quad N_{CD} = -R_D = -3(71.3) = \boxed{-213.9 \text{ kN}}$$
(compression).

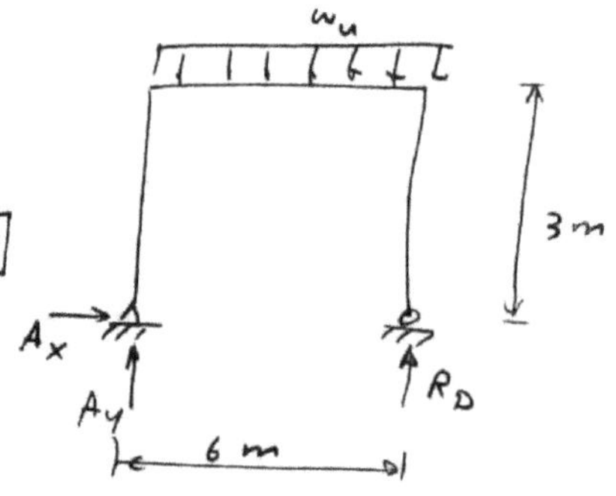

<u>Case 2</u>: $L+D+W$: (W acts to the right):
There is <u>no</u> symmetry in this case.

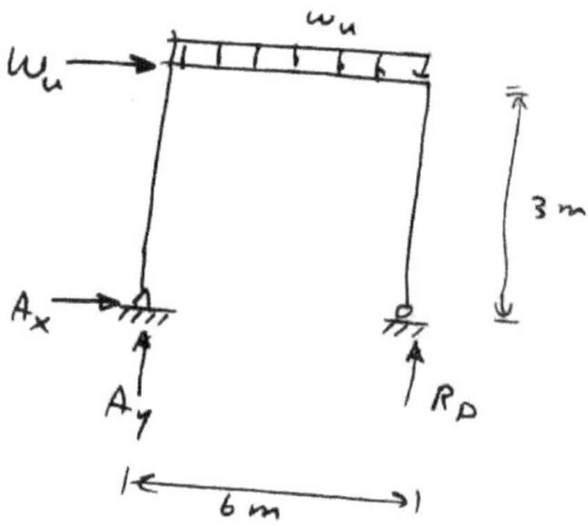

$$w_u = 0.75\,(1.4D + 1.7L)$$
$$= 0.75\,[1.4(14.5) + 1.7(30)]$$
$$= 53.475 \text{ kN/m}.$$

$$W_u = 0.75\,(1.7W)$$
$$= 0.75\,[1.7(90)] = 114.75 \text{ kN}.$$

$+\!\!\curvearrowright \Sigma M_A = 0$:

$$R_D(6) - W_u(3) - w_u(6)(3) = 0$$

$$R_D(6) - 114.75(3) - 53.475(6)(3) = 0$$

$$\therefore \quad R_D = 217.8 \text{ kN}.$$

$$\therefore \quad N_{CD} = -R_D = \boxed{-217.8 \quad \text{kN}} \quad (\text{compression}).$$

<u>Case 3</u>: $L+D+W$ (W acts to the left):
There is <u>no</u> symmetry in this case.

$$w_u = 53.475 \text{ kN/m}.$$
$$W_u = 114.75 \text{ kN}.$$

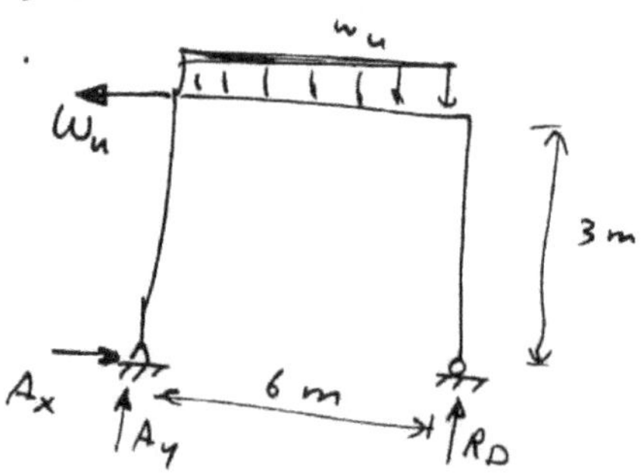

$+\!\!\curvearrowright \Sigma M_A = 0$:

$$R_D(6) + 114.75(3) - 53.475(6)(3) = 0$$

$$\therefore \quad R_D = 103 \text{ kN}.$$

$$\therefore \ N_{CD} = -R_D = \boxed{-103 \text{ kN}} \quad (\text{compression}).$$

Case 4 : <u>D + W only</u> (W acts to the right) :

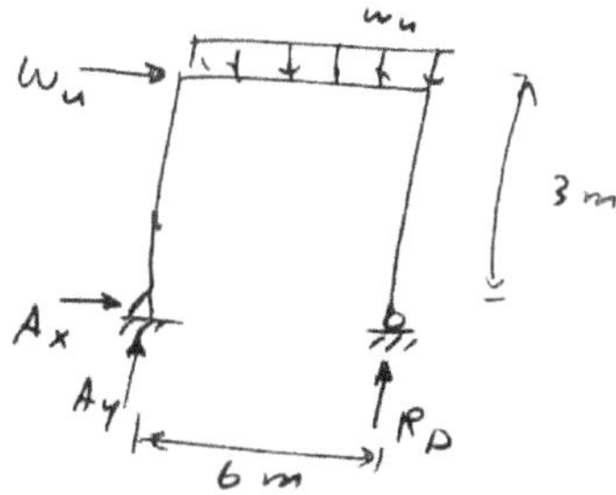

$$w_u = 0.9 \, D = 0.9 \,(14.5) = 13.05 \text{ kN/m}.$$
$$W_u = 1.3 \, W = 1.3 \,(90) = 117 \text{ kN}.$$

$+\circlearrowright \ \Sigma M_A = 0 :$

$$R_D(6) - 117(3) - 13.05 \,(6)(3) = 0$$
$$\therefore \ R_D = +97.65 \text{ kN}.$$
$$\therefore \ N_{CD} = -R_D = \boxed{-97.65 \text{ kN}} \quad (\text{compression}).$$

Case 5 : <u>D + W only</u> (W acts to the left) :

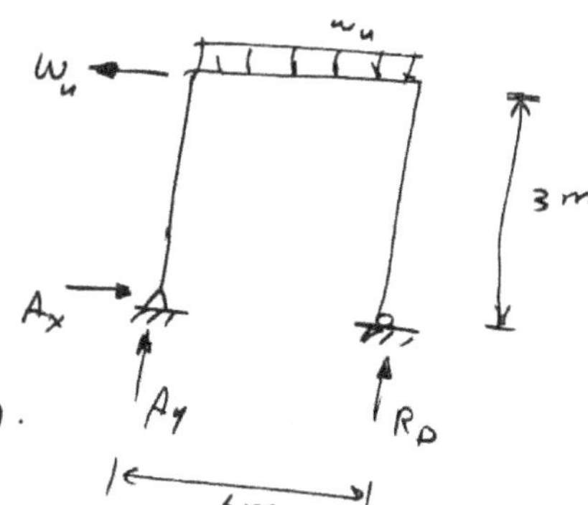

$$w_u = 13.05 \text{ kN/m}.$$
$$W_u = 117 \text{ kN}.$$

$+\circlearrowright \ \Sigma M_A = 0 :$

$$R_D(6) + 117(3) - 13.05 \,(6)(3) = 0$$
$$R_D = -19.35 \text{ kN}.$$
$$\therefore \ N_{CD} = -R_D = \boxed{+19.35 \text{ kN}} \quad (\text{tension}).$$

$\therefore$ the two most severe cases are
Cases (2) and (5).

$\therefore$ Member CD must be designed with a compressive
strength of 213.975 kN and a tensile strength of 19.35 kN

Example 2 :

The beam in the figure carries service loads that consist of :

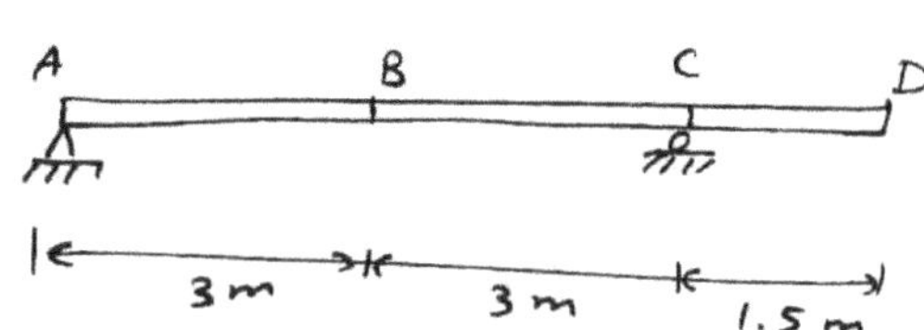

(a) a uniformly distributed dead load $w_d = 20$ kN/m acting over its entire length, and

(b) a concentrated live load $P_\ell = 105$ kN that can act anywhere on the span.

Using the ACI load factors and reduction factors, determine the required nominal flexural strength M_n at both point B and at point C.

Solution : We have only one case of load combination, i.e. $D+L$ only.

$$\omega_u = 1.4\,\omega_d = 1.4\,(20) = 28 \text{ kN/m}.$$

$$P_u = 1.7\,P_\ell = 1.7\,(105) = 178.5 \text{ kN}.$$

There are two position for the line load P_u that produce maximum moments at points B and C.

Position 1 : Position the line load at B:

$\xrightarrow{+} \Sigma F_x = 0: \quad A_x = 0 .$

$\circlearrowright^+ \Sigma M_A = 0:$

$\quad C_y (6) - 178.5 (3)$

$\quad - 28 (7.5) \left(\dfrac{7.5}{2}\right) = 0$

$\quad \therefore C_y = 220.5 \text{ kN} \uparrow .$

$+\uparrow \Sigma F_y = 0:$

$\quad A_y + 220.5 - 178.5 - 28(7.5) = 0$

$\quad \therefore A_y = 168 \text{ kN} \uparrow .$

At point B :

$\quad M_u = 378 \text{ kN} \cdot m$

$\quad M_u = \phi M_n$

but $\phi = 0.9$ for flexure.

$\quad \therefore 378 = 0.9\, M_n$

$\quad \therefore M_n = \boxed{420 \text{ kN} \cdot m}$

At point C :

$\quad M_u = -31.5 \text{ kN} \cdot m$

$\quad M_u = 0.9\, M_n$

$\quad -31.5 = 0.9\, M_n$

$\quad \Rightarrow M_n = \boxed{-35 \text{ kN} \cdot m}$

Position 2 : Position the line load at D :

$\xrightarrow{+} \Sigma F_x = 0: \quad A_x = 0 .$

$\circlearrowright^+ \Sigma M_A = 0: \quad C_y(6) - 178.5(7.5) - 28(7.5)\left(\dfrac{7.5}{2}\right) = 0$

$\quad \therefore C_y = 354.375 \text{ kN} \uparrow .$

$+\uparrow \Sigma F_y = 0: \quad A_y + 354.375 - 178.5 - 28(7.5) = 0$

$\quad \therefore A_y = 34.125 \text{ kN} \uparrow .$

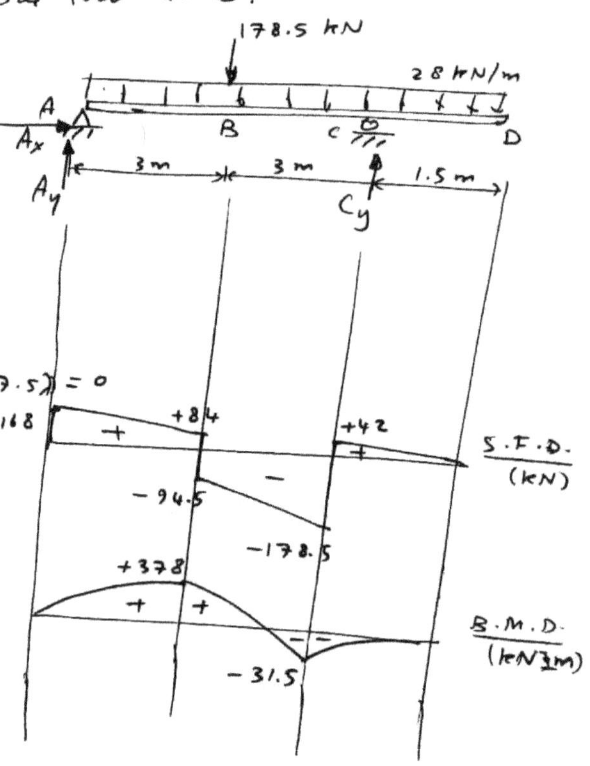

Locate the point O of zero shear:

$$\frac{x}{6-x} = \frac{34.125}{133.875}$$

$$133.875\,x = 6\,(34.125) - 34.125\,x$$

$$\therefore\ x = 1.22\ m\ .$$

$$\therefore\ m_{o_u} = \frac{1}{2}(34.125)(1.22)$$

$$= 20.795\ kN\cdot m$$

At point B :

$$\frac{V_B}{3-1.22} = \frac{34.125}{1.22}$$

$$\therefore\ V_B = 49.79\ kN\ .$$

$$\therefore\ M_B = 20.795 - \frac{1}{2}(49.79)(3-1.22)$$

$$= -23.518\ kN\cdot m\ .$$

$$\therefore\ M_u = \phi M_n$$

$$-23.518 = 0.9\,M_n$$

$$\therefore\ M_n = \boxed{-26.13\ kN\cdot m}$$

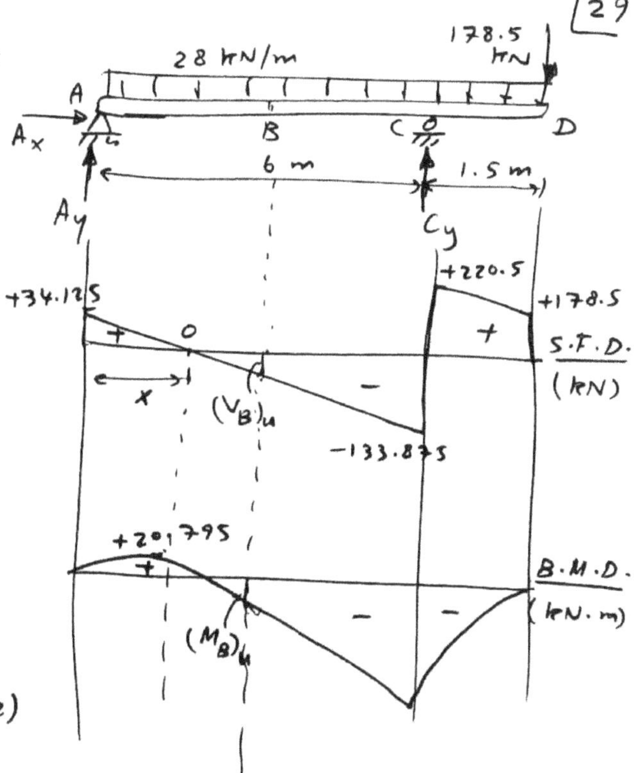

At point C :

$$M_c = -23.518 - \frac{1}{2}(49.79 + 133.875)(3) = -299.02\ kN\cdot m\ .$$

$$M_u = \phi M_n$$

$$-299.02 = 0.9\,M_n \implies M_n = \boxed{-332.24\ kN\cdot m}$$

<u>Results</u> : * the moment at B changes sign from $M_n = 420$ kN·m to $M_n = -26.13$ kN·m. The beam must be designed to carry both positive and negative moments. Thus, we must reinforce both the top and bottom of the cross-section with longitudinal steel.

Both Positions (1) & (2) are critical for M_B.

* the moment at C is longer for Position (2) with $M_n = -332.24$ kN·m. (<u>required nominal flexural strength</u>)

Example 3:

Determine the minimum cross-sectional area A of a short steel column required to support a live load $L = 133$ kN and a dead load $D = 178$ kN. The figure shows the column with the stress-strain curve for steel. Failure is assumed to occur when the average stress on the cross-section reaches the yield point stress f_y. Use a reduction factor $\phi = 0.7$ ($f_y = 345$ MPa).

Solution:

$$P_u = 1.4D + 1.7L = 1.4(178) + 1.7(133)$$
$$= 475.3 \text{ kN}.$$

Calculate the design strength ϕP_n:

$$\phi P_n = \phi(A f_y) \quad \text{where} \quad f_y = \frac{P_n}{A}$$

$$= 0.7 A (345) \times 10^3$$

$$\phi P_n = 241.5 A \times 10^3$$

Required Strength $\leq$ Design Strength

$$P_u \leq \phi P_n$$

Take $P_u = \phi P_n$

$$475.3 = 241.5 A \times 10^3 \implies A = 1.96812 \times 10^{-3} \text{ m}^2$$
$$= 1.96812 \times 10^{-3} \times (10^3)^2$$
$$= 1968.12 \text{ mm}^2$$
$$= 19.6812 \text{ cm}^2.$$

$f_y = 345$ MPa
$= 345 \times 10^3$ kN/m²

Finally, check for service loads:

when the full service load $P = L + D = 133 + 178 = 311$ kN acts,

we have: elastic stress $f \equiv \sigma = \dfrac{P}{A} = \dfrac{311}{1.96812 \times 10^{-3}} = 158018.82$ kN/m²

$$= 158 \text{ MPa}$$
$$< f_y = 345 \text{ MPa}.$$

2.9 Ductility vs. Brittleness:

* Plain concrete is <u>brittle</u> but steel is <u>ductile</u>.

* The term <u>ductility</u> describes the ability of a member to undergo large deformations without rupture as failure occurs.

* Ductile structures may bend and deform excessively under load, by they remain intact. This capability prevents total collapse and provides protection to occupants of buildings.

* The term <u>brittle</u> describes members that fail suddenly, completely, and with little warning.

* When a brittle member fractures, it usually disintegrates and may damage adjacent portions of the structure or overload other members, bringing on additional failures.

* The ability of a <u>ductile structure</u> to undergo large deformations before collapse produces visible evidence of impending failure and may give occupants the opportunity to relieve distress by reducing loads.

* In contrast, <u>brittle failures</u> occur suddenly, without warning, and with no time for measures to be taken to prevent damage.

* A major objective of the ACI Code is design concrete structures with adequate ductility since concrete is brittle without reinforcement.

2.10 Strength and Serviceability:

* Although it is imperative that structures be designed with adequate strength to reduce the probability of failure to an acceptable level, they must also function effectively under service loads. The following <u>serviceability</u> requirements must be considered:

1. Deflections must be limited to ensure that floors will remain level within required tolerances,
2. No vibrations should occur excessively,

3. Cracking must be prevented especially in plaster ceilings and masonry partitions.

4. Sensitive equipment will not be thrown out of alignment.

5. The width of cracks must be limited to preserve the architectural appearance of exposed surfaces and to protect reinforcement from corrosion.

2.11 Accuracy of Computations:

* Design calculations for reinforced concrete structures do <u>not</u> require a high degree of precision because variations exist in material strength for both steel and concrete, and variations in dimensions are inevitable (and acceptable).

* Reinforced concrete design is <u>not</u> an exact science for which answers can be confidentally calculated to six or eight places.

* Recording of results on computation sheets should <u>not</u> exceed four significant figures, primarily for systematic control and checking of computations.

* The designer should place highest priority on determining proper location and length of steel reinforcement to carry the tension forces, thus making up for that capacity which is deficient in the concrete.

* Failures, when they occur, generally result from gross underestimating of tensile forces, or lack of identification of how the structure or element will behave under loads.

* Failures are rarely the result of carrying too few significant figures in the design computations. However, significant figures may be lost in arithmetic operations, and gross errors may sometimes result from sloppiness.

Chapter 3
Strength of Rectangular Sections in Bending
Part -I-

3.1 General Introduction:

* Since 1971, the ACI Code used the strength method of design.

* In the strength design method, factored loads and computed strength of sections at imminent failure are used.

(1) <u>Factored loads</u> (including moments, shear, axial forces, etc) are obtained by multiplying the service loads by load factors to account for possible variations in design assumptions and overloads.

(2) The <u>design strength</u> of a section is obtained by multiplying the nominal strength by a strength <u>reduction factor</u> ϕ to account for adverse variations in material strengths, workmanship, dimensions, control, and degree of supervision.

3.2 Basis of Nominal Flexural Strength, M_n :

* F. Stüssi (1932) started the reinforced concrete beam strength design.

* The stress-strain curve is <u>nonlinear</u> for levels above $0.5 f_c'$.

* In fact, this variation is shown in the figure below.

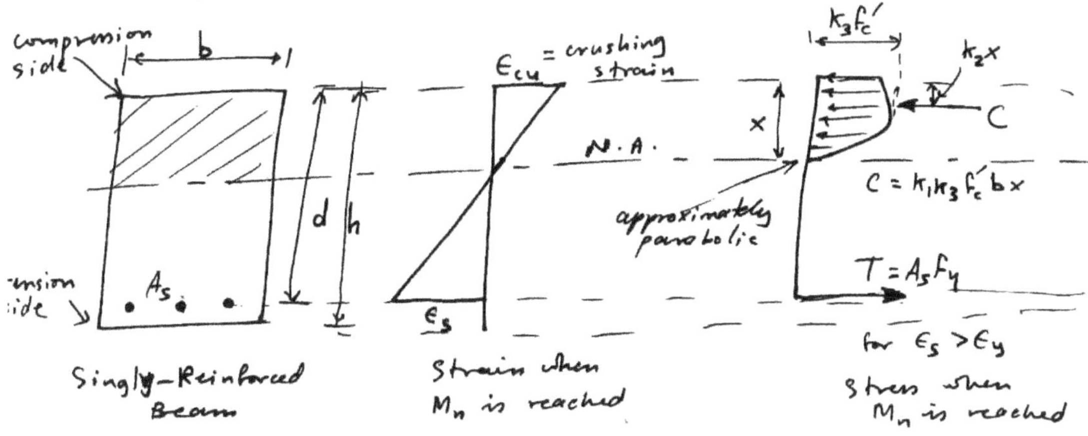

* The nominal strength is assumed to be reached when the strain in the extreme compression fiber is equal to the crushing strain ϵ_{cu} of concrete.

* When crushing occurs (a somewhat sudden occurrence), the strain in the tension steel A_s <u>could be</u> either larger or smaller than the yield strain ϵ_y which is obtained from:

$$f_y = E_s \epsilon_y \implies \epsilon_y = \frac{f_y}{E_s}.$$

* There are two cases:

(1) If the steel amount is <u>low</u> enough, it would yield prior to crushing of the concrete, resulting in a <u>ductile failure</u> mode in which there is large deformation.

(2) If the amount of steel is <u>high</u> enough, it would allow the steel to remain elastic at the time of crushing of the concrete, causing a <u>brittle</u> or <u>sudden</u> mode of <u>failure</u>.

* The ACI Code provides requirements on the amount of steel to ensure the ductile mode of failure when the nominal strength is reached.

* The (maximum) compressive stress of concrete in a beam is actually less than f'_c (the compressive strength in the lab), it is actually $k_3 f'_c$.

* The average stress over a beam of constant width is $k_1 k_3 f'_c$.

* The centroidal location of the (roughly) parabolic distribution is $k_2 x$, measured from the compressive face.

* x : neutral axis distance.

$$\therefore \quad \text{Force} = \text{Stress} \times \text{Area}$$
$$C = (k_1 k_3 f'_c)(bx) = k_1 k_3 f'_c \, x \, b \quad \text{(total compressive force)}.$$

* Assuming that ductile failure occurs first (i.e. steel yields),

$$\therefore \quad T = f_y A_s \qquad \text{(stress} \times \text{area)}$$

* Equation of Equilibrium: $\overset{+}{\rightarrow} \Sigma F_x = 0 : \quad T - C = 0$

$$\therefore \quad C = T$$
$$k_1 k_3 f'_c \, bx = f_y A_s$$

$$\therefore \quad x = \frac{f_y A_s}{k_1 k_3 f_c' b}$$

$M_n \equiv$ nominal flexural strength

$\quad = $ moment about C

$\quad = T(d - k_2 x)$

$\quad = f_y A_s (d - k_2 x)$

$\quad = f_y A_s \left(d - k_2 \dfrac{f_y A_s}{k_1 k_3 f_c' b}\right)$

$$\boxed{M_n = f_y A_s \left(d - \frac{k_2}{k_1 k_3} \frac{A_s f_y}{f_c' b}\right)} \quad \underline{\text{Nominal Flexural Strength}}$$

* M_n can be determined if $\dfrac{k_2}{k_1 k_3}$ is known.

* $\dfrac{k_2}{k_1 k_3}$ values are obtained experimentally. They range between $0.55 \rightarrow 0.63$. (<u>Note</u>: they depend on the value of f_c').
(see the figure).

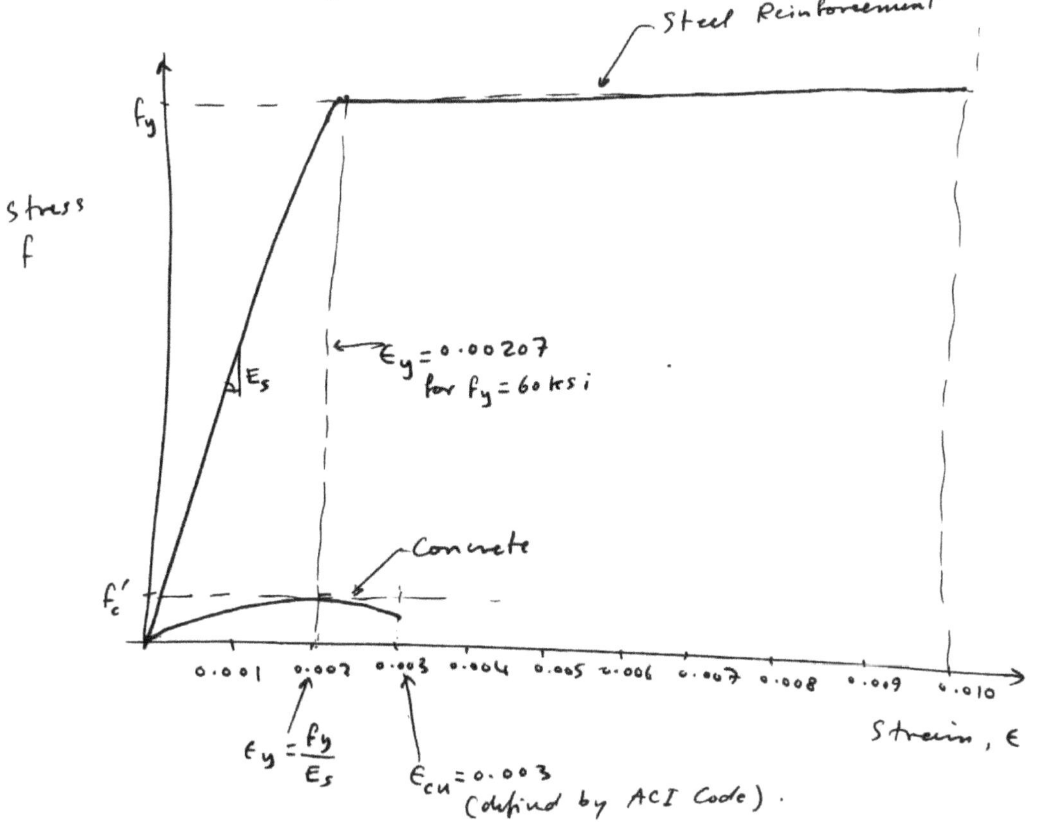

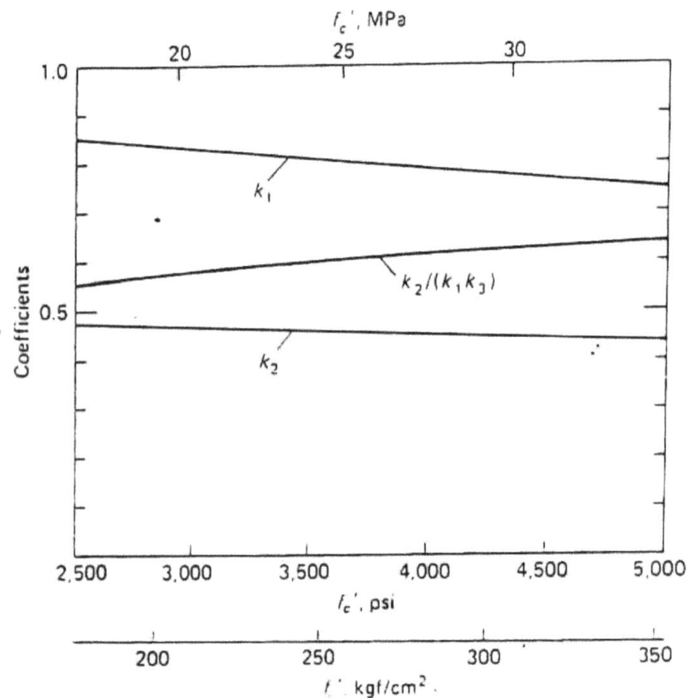

Figure 3.2.3 "Stress solid" parameters (adapted from Ref. 3.2)

TNEY RECTANGULAR STRESS DISTRIBUTION

The computation of flexural strength M_n based on the approximately parabolic stress distribution of Fig. 3.2.1(c) may be done using Eq. (3.2.5) with given values of $k_2/(k_1 k_3)$. However, it is desirable for the designer to have a simple method in which basic static equilibrium is used.

In the 1930s, Whitney [3.5, 3.6] proposed the use of a rectangular compressive stress distribution to replace that of Fig. 3.2.1(c). As shown in Fig. 3.3.1(c), an average stress of $0.85f_c'$ is used with a rectangle of depth $a = \beta_1 x$, determined so that $a/2 = k_2 x$. Whitney determined that β_1 should be 0.85 for concrete with $f_c' \leq 4000$ psi, and 0.05 less for each 1000 psi of f_c' in excess of 4000 psi.* The value of β_1 may not be taken less than 0.65 (ACI-10.2.7.3).

The flexural strength M_n, using the equivalent rectangle, is obtained from Fig. 3.3.1 as follows:

$$C = 0.85f_c'ba \tag{3.3.1}$$

$$T = A_s f_y \tag{3.3.2}$$

---

* For SI, ACI 318–89M gives $\beta_1 = 0.85$ for $f_c' \leq 30$ MPa and reduces by 0.08 for each 10 MPa of f_c' in excess of 30 MPa, but not less than 0.65.

* For the case when crushing of concrete occurs, the ACI Code considers the maximum crushing strain $\underline{\epsilon_{cu} = 0.003}$.

3.3 Whitney Rectangular Stress Distribution:

* Whitney (in the 1930s) proposed the use of a rectangular compressive stress distribution to replace the approximately parabolic stress distribution.

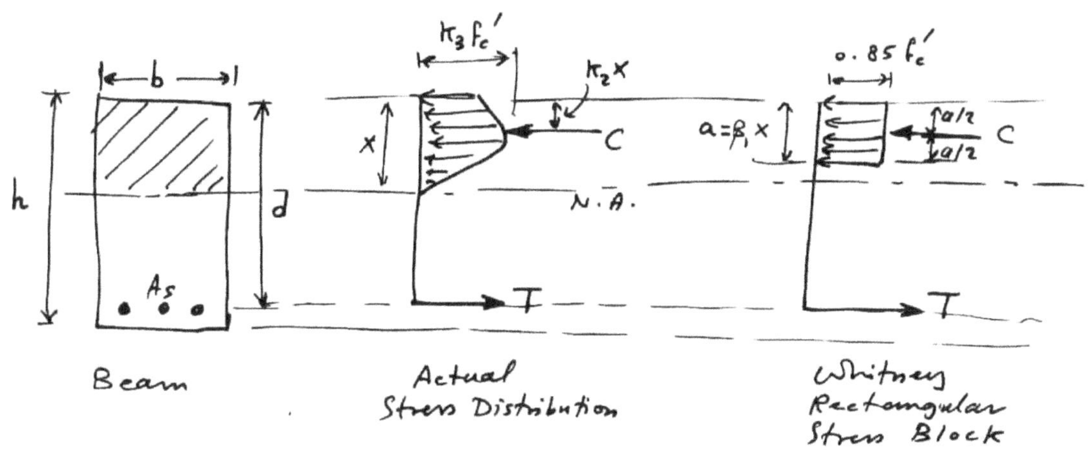

* An average stress of $0.85 f_c'$ is assumed (instead of $k_1 k_3 f_c'$).
  (i.e. take $k_1 k_3 = 0.85$).

* Assume a rectangular stress block of depth $a = \beta_1 x$ determined so that $\dfrac{a}{2} = k_2 x$.

* Whitney determined that β_1 should be:

$$\beta_1 = \begin{cases} 0.85 & \text{for concrete with } f_c' \leq 4000 \, psi \\ 0.05 \text{ less for each } 1000 \, psi \text{ of } f_c' \text{ in excess of } 4000 \, psi. \end{cases}$$

  $\underline{but}$ not less than 0.65 (ACI Code, Section 10.2.7.3).

  In SI Units,

$$\beta_1 = \begin{cases} 0.85 & \text{for concrete with } f_c' \leq 30 \, MPa \\ 0.08 \text{ less for each } 10 \, MPa \text{ of } f_c' \text{ in excess of } 30 \, MPa \end{cases}$$

  $\underline{but}$ not less than 0.65.

$\therefore$ Force = Stress × Area

$$C = 0.85 f_c' \, a b$$

$$T = f_y A_s \quad \text{(assuming steel yields prior to crushing of concrete)}$$

$\therefore$ $C = T \implies 0.85 f_c' \, a b = f_y A_s \implies a = \dfrac{f_y A_s}{0.85 f_c' b}$

Take moment about C: $M_n = T \left(d - \dfrac{a}{2} \right)$

$$= f_y A_s \left(d - \dfrac{f_y A_s}{2(0.85 f_c' b)} \right)$$

$$\boxed{\therefore \; M_n = f_y A_s \left(d - 0.59 \dfrac{f_y A_s}{f_c' b} \right)} \quad \begin{array}{l}\text{Nominal}\\ \text{Flexural}\\ \text{Strength}\end{array}$$

Notice that $\dfrac{k_2}{k_1 k_3} = 0.59$ here (average value of 0.55 and 0.63).

* ACI Code accepts the Whitney rectangle (Section 10.2.7).

* β_1 is needed only to establish the neutral axis location for determining the steel strain.

* As long as the steel strain exceeds $\varepsilon_y = \dfrac{f_y}{E_s}$ (steel yielding), the flexural strength M_n is not affected by the value of β_1.

3.4 Rectangular Sections in Bending with Tension Reinforcement Only:

* A section is said to be <u>singly reinforced</u> if it contains tension reinforcement only.

* The quantities defining a singly reinforced rectangular section are: $\begin{Bmatrix} 1. & b \\ 2. & d \\ 3. & A_s \end{Bmatrix}$

* Protective concrete cover is necessary around the steel bars in order to: 1. make steel and concrete act together   2. provide fire protection for steel.

* At high temperatures, f_y and E_s (for steel) begin to reduce so that concrete cover is needed for insulation. (See ACI-Code, Section 7.7).

* ACI Code (Section 10.2.5) neglects the tensile strength of concrete in flexure calculations.
* Therefore, the cross-sectional shape of the beam on the tension side of the neutral axis and the amount of concrete cover do not affect the flexural strength, M_n.

* thus, d (effective depth) is the important depth dimension for computing strength.
* the <u>effective depth, d,</u> is defined as the distance from the extreme fiber in compression to the centroid of the tension steel area.

<u>ACI Code Strength Method:</u>

1. Calculate the factored loads $U = 1.4D + 1.7L$ (for D & L only).
   <u>or</u> Calculate the factored moments,
   <u>i.e.</u> $\qquad M_u = 1.4 M_D + 1.7 M_L$.

2. Required flexural strength $\quad M_u \leq \phi M_n$
   where $\phi = 0.90$ for flexure.

* For computation of the nominal flexural strength M_n, the following assumptions (ACI Code, Section 10.2) are used:

(1) the strength of members shall be based on satisfying the applicable conditions of equilibrium and compatibility of strains.

(2) Strain in the steel reinforcement and the concrete shall be assumed directly proportional to the distance from the neutral axis (except for deep members, ACI – Sec. 10.7).

(3) The maximum usable (crushing) strain at the extreme concrete compression fiber shall be assumed to equal to 0.003.

(4) The tensile strength of the concrete is to be neglected.

(5) the modulus of elasticity of nonprestressed steel reinforcement may be taken as 29,000,000 psi (200,000 MPa) or 2,040,000 kg/cm²).

(6) the relationship between the concrete compressive stress distribution and the concrete strain when nominal strength

is reached, may be taken as an equivalent rectangular stress distribution (ACI Code, Section 10.2.7), wherein a concrete stress intensity of $0.85 f_c'$ is assumed to be uniformly distributed over an equivalent compression zone bounded by the edges of the cross-section and a straight line located parallel to the neutral axis at a distance $a = \beta_1 x$ from the fiber of maximum compressive strain.

* The value of β_1 is given by:

$$\beta_1 = \begin{cases} 0.85 & , \text{ for } f_c' \le 4000 \text{ psi} \\ 0.85 - 0.05\left(\dfrac{f_c' - 4000}{1000}\right) \ge 0.65 & , \text{ for } f_c' > 4000 \text{ psi} \end{cases}$$

<u>In SI units,</u>

$$\beta_1 = \begin{cases} 0.85 & , \text{ for } f_c' \le 30 \text{ MPa} \\ 0.85 - 0.08\left(\dfrac{f_c' - 30}{10}\right) \ge 0.65 & , \text{ for } f_c' > 30 \text{ MPa} \end{cases}$$

<u>Example 1</u>:

Determine the nominal flexural strength M_n of the rectangular section shown in the figure, given $f_c' = 35$ MPa, $f_y = 350$ MPa, $b = 350$ mm, $d = 550$ mm, and $A_s = 4 - 30$ mm bars.

<u>Solution</u>:

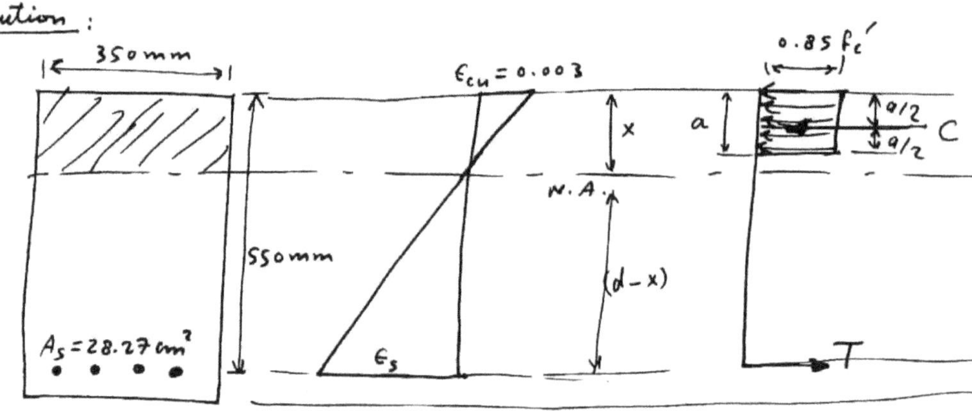

$$A_s = 4\left(\frac{\pi d^2}{4}\right) = 4\left(\frac{\pi (3)^2}{4}\right) = 28.27 \text{ cm}^2 \quad (\text{or from tables}).$$

$$= 28.27 \times 10^2 \text{ mm}^2$$
$$= 2827 \text{ mm}^2.$$

Assume that the steel has yielded when the strength is reached.

$$C = 0.85 f_c' a b = 0.85 (35)(350) a = 10412.5 a$$

$$T = f_y A_s = (350)(2827) = 989,450$$

$$C = T \implies 10412.5 a = 989450 \implies a = 95 \text{ mm}.$$

For $f_c' = 35 \text{ MPa}$, we have $\beta_1 = 0.85 - 0.08 \left(\frac{35 - 30}{10} \right) = 0.81$.

The neutral axis position is determined by $a = \beta_1 x$.

$$\therefore \quad x = \frac{a}{\beta_1} = \frac{95}{0.81} = 117.3 \text{ mm}.$$

* The flexural nominal strength M_n is calculated by:

$$M_n = C \left(d - \frac{a}{2} \right) = T \left(d - \frac{a}{2} \right)$$

$$= 0.85 (35 \times 1000) \left(\frac{350}{1000} \right) \left(\frac{95}{1000} \right) \left(\frac{550}{1000} - \frac{95}{2(1000)} \right)$$

$$= 497 \text{ kN} \cdot \text{m}.$$

<u>Check</u> that steel yields first:

From similar triangles, $\dfrac{x}{\epsilon_{cu}} = \dfrac{d - x}{\epsilon_s}$

we assume $\epsilon_{cu} = 0.003$.

$$\therefore \quad \frac{117.3}{0.003} = \frac{550 - 117.3}{\epsilon_s} \implies \epsilon_s = 0.0111$$

The steel yield strain $\epsilon_y = \dfrac{f_y}{E_s} = \dfrac{350}{200,000} = 0.00175$.

$$\therefore \quad \epsilon_s = 0.0111 > 0.00175 \implies \therefore \text{ steel has yielded first.}$$

<u>Example 2</u>:

For the beam of Example 1, determine the safe service moment M_w that may be applied according to the ACI Code, if 60% of the total moment is dead load and 40% is live load.

__Solution:__ $M_n = 497 \ kN \cdot m$ (nominal strength).

$M_u = 1.4 \, M_D + 1.7 \, M_L$ (factored moment).

$M_u \leq \phi \, M_n$ for safety, with $\phi = 0.9$ for flexure

but we are given that $M_D = 60\% \, M_w = 0.6 \, M_w$
$M_L = 40\% \, M_w = 0.4 \, M_w$.

$\therefore \quad M_u = 1.4 \, M_D + 1.7 \, M_L = 1.4 \, (0.6 \, M_w) + 1.7 \, (0.4 \, M_w)$
$= 1.52 \, M_w$.

$\therefore \quad M_u \leq \phi \, M_n$
$1.52 \, M_w \leq 0.9 \, (497) \implies M_w \leq \underline{\underline{294}} \ kN \cdot m$.

$\therefore \quad \text{Factor of Safety} = \dfrac{M_n}{M_w} = \dfrac{497}{294} = 1.69$.

3.5 Definition of Balanced Strain Condition:

* At the __balanced strain condition__, the maximum strain ϵ_{cu} of the extreme concrete compression fiber just reaches 0.003 __simultaneously__ with the tension steel reaching a strain $\epsilon_y = \dfrac{f_y}{E_s}$.

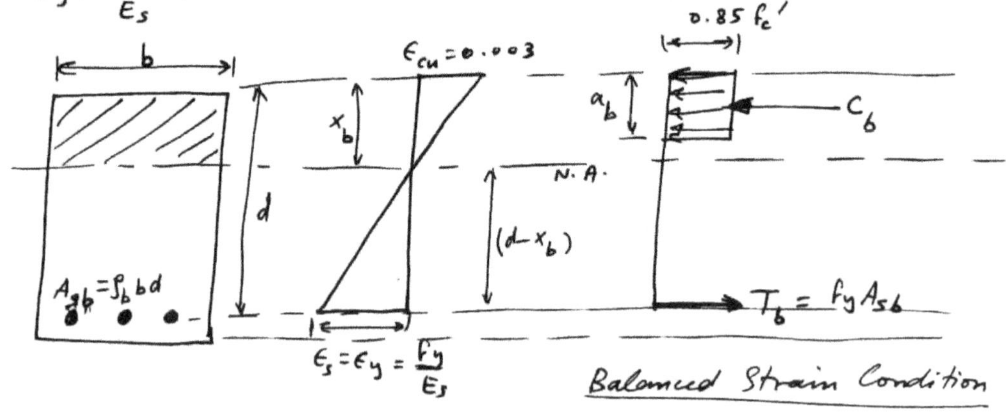

A_{sb}: balanced steel area $= \rho_b \, bd$

__Case 1:__ If $A_s > A_{sb}$, then $a > a_b$, $x > x_b$
and $\epsilon_s < \epsilon_y = \dfrac{f_y}{E_s}$ when $\epsilon_{cu} = 0.003$

$\therefore$ Crushing of concrete occurs first (undesirable).
Failure of the beam will be sudden with no warning.

<u>Case 2</u>: If $A_s < A_{sb}$, $\quad x < x_b$, $\quad a < a_b$,

$$\therefore \quad \epsilon_s = \epsilon_y = \frac{f_y}{E_s} \quad , \text{ steel yields first },$$

the beam will have noticeable deflection prior to the concrete reaching the crushing strain $\epsilon_{cu} = 0.003$ (gradual failure).

*** Reinforcement Ratio ρ_b at Balanced Strain Condition for Rectangular Beams Having Tension Reinforcement Only:**

ρ : reinforcement ratio (reinforcement percentage) (percentage of steel)

$$\rho \equiv \frac{A_s}{bd} = \frac{\text{area of steel}}{\text{effective area}}$$

ρ_b : reinforcement ratio at the balanced strain condition.

By similar triangles, $\quad \dfrac{x_b}{\epsilon_{cu}} = \dfrac{d - x_b}{\epsilon_y}$

$$\therefore \quad \epsilon_{cu}(d - x_b) = \epsilon_y \, x_b$$

$$\epsilon_{cu} d - \epsilon_{cu} x_b = \epsilon_y \, x_b$$

$$\epsilon_{cu} d = (\epsilon_{cu} + \epsilon_y) x_b \quad \Rightarrow \quad \frac{x_b}{d} = \frac{\epsilon_{cu}}{\epsilon_{cu} + \epsilon_y}$$

$$\therefore \quad \frac{x_b}{d} = \frac{\epsilon_{cu}}{\epsilon_{cu} + \dfrac{f_y}{E_s}} = \frac{0.003}{0.003 + \dfrac{f_y}{200,000}} \quad , \quad f_y \text{ in MPa}$$

$$\therefore \quad \frac{x_b}{d} = \frac{600}{600 + f_y}$$

the compressive balanced force $\quad C_b = 0.85 f_c' \, b \, a_b$, but $a_b = \beta_1 x_b$

$$\therefore \quad C_b = 0.85 f_c' b \beta_1 x_b$$

the tensile balanced force $\quad T_b = f_y A_{sb} = f_y \cdot \rho_b \, bd$

$$\text{where} \quad \rho_b = \frac{A_{sb}}{bd}$$

Equilibrium $\Rightarrow$ $\qquad C_b = T_b$

$$0.85\, f_c'\, b\, \beta_1 X_b = f_y\, \rho_b\, b\, d$$

$$\therefore\ 0.85\, f_c'\, b\, \beta_1 \cdot \left(\frac{600}{600 + f_y}\right) d = f_y\, \rho_b\, b\, d$$

bd cancels $\Rightarrow$ $\qquad 0.85\, f_c'\, \beta_1 \left(\frac{600}{600 + f_y}\right) = \rho_b\, f_y$

$$\therefore\ \boxed{\rho_b = \frac{0.85\, f_c'}{f_y}\, \beta_1 \left(\frac{600}{600 + f_y}\right)}\quad \text{where}\quad f_y, f_c'\ \text{are in MPa.}$$

In English units, $\qquad \boxed{\rho_b = \frac{0.85\, f_c'}{f_y}\, \beta_1 \left(\frac{87000}{87000 + f_y}\right)}$ where f_y, f_c' are in psi.

* The balanced strain condition is <u>not</u> allowed to occur by ACI Code.

3.6 Maximum Reinforcement Ratio ρ :

* ACI Code (Section 10.3.3) limits the amount of tension steel to not more than 75% of the amount in the balanced strain condition.

* This is done to ensure a ductile mode of failure in flexure.

$$\rho_{max.} = 0.75\, \rho_b$$

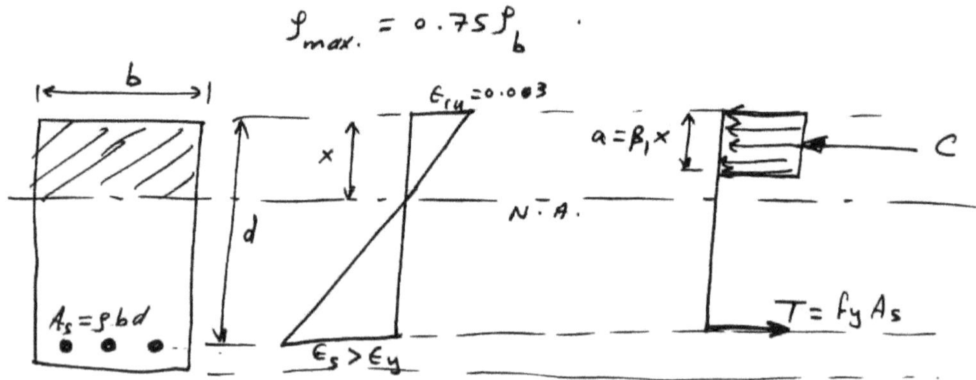

The above formula is equivalent to $\quad X_{max} = 0.75\, X_b$.
(for rectangular beams, singly reinforced).

Example 3 :

Determine whether or not the amount of steel used in the beam ($b = 350$ mm, $d = 550$ mm, $A_s = 2827$ mm^2, $f_c' = 35$ MPa, and $f_y = 350$ MPa) of Example 1 is acceptable according to the ACI Code ?

Solution:

Determine A_{sb} in the balanced strain condition.

From similar triangles,

$$\frac{x_b}{\epsilon_{cu}} = \frac{d - x_b}{\epsilon_y}$$

$$\epsilon_y x_b = \epsilon_{cu} d - \epsilon_{cu} x_b$$

$$(\epsilon_y + \epsilon_{cu}) x_b = \epsilon_{cu} d$$

$$\therefore \quad x_b = \left(\frac{\epsilon_{cu}}{\epsilon_{cu} + \epsilon_y}\right) d$$

$$\therefore \quad x_b = \left(\frac{\epsilon_{cu}}{\epsilon_{cu} + \frac{f_y}{E_s}}\right) d = \left(\frac{0.003}{0.003 + \frac{350}{200,000}}\right)(550) = 347.4 \text{ mm}.$$

For $f_c' = 35 \text{ MPa}$, we have $\beta_1 = 0.81$

$$\therefore \quad a_b = \beta_1 x_b = (0.81)(347.4) = 281.4 \text{ mm}.$$

$$C_b = 0.85 f_c' a_b b = T_b = A_{sb} f_y$$

$$\therefore \quad A_{sb} = \frac{0.85 f_c' a_b b}{f_y} = \frac{0.85\left(\frac{35}{350}\right)(281.4)(350)}{} $$
$$= 8372 \text{ mm}^2.$$

$$A_{s_{max.}} = 0.75 A_{sb} = 0.75(8372) = 6279 \text{ mm}^2.$$

$$A_s = 2827 \text{ mm}^2 < 6279 \text{ mm}^2.$$

$$\therefore \quad \text{ACI Code (Section 10.3.3) is satisfied.} \quad \checkmark$$

$$\rho_b = \frac{A_{sb}}{bd} = \frac{8372}{350 \times 550} = 0.0435 = 4.35\%.$$

$$\rho_{max.} = 0.75 \frac{A_{sb}}{bd} = 3.26\%.$$

$$\text{Actual } \rho = \frac{A_s}{bd} = \frac{2827}{350 \times 550} = 1.47\%.$$

Table 3.6.1   Maximum Reinforcement Ratio ρ for Singly Reinforced Rectangular Beams (Corresponding to $0.75\rho_b$)

f_y	$f'_c = 3000$ psi $\beta_1 = 0.85$	$f'_c = 3500$ psi $\beta_1 = 0.85$	$f'_c = 4000$ psi $\beta_1 = 0.85$	$f'_c = 5000$ psi $\beta_1 = 0.80$	$f'_c = 6000$ psi $\beta_1 = 0.75$
40,000 psi	0.0278	0.0325	0.0371	0.0437	0.0491
50,000 psi	0.0206	0.0241	0.0275	0.0324	0.0364
60,000 psi	0.0160	0.0187	0.0214	0.0252	0.0283

f_y	$f'_c = 20$ MPa $\beta_1 = 0.85$	$f'_c = 25$ MPa $\beta_1 = 0.85$	$f'_c = 30$ MPa $\beta_1 = 0.85$	$f'_c = 35$ MPa $\beta_1 = 0.81$	$f'_c = 40$ MPa $\beta_1 = 0.77$
300 MPa	0.0241	0.0301	0.0361	0.0402	0.0436
350 MPa	0.0196	0.0244	0.0293	0.0326	0.0354
400 MPa	0.0163	0.0203	0.0244	0.0271	0.0295

f_y	$f'_c = 200$ kgf/cm^2 $\beta_1 = 0.85$	$f'_c = 240$ kgf/cm^2 $\beta_1 = 0.85$	$f'_c = 280$ kgf/cm^2 $\beta_1 = 0.85$	$f'_c = 320$ kgf/cm^2 $\beta_1 = 0.82$	$f'_c = 360$ kgf/cm^2 $\beta_1 = 0.79$
2800 kgf/cm^2	0.0266	0.0319	0.0372	0.0410	0.0444
3500 kgf/cm^2	0.0197	0.0236	0.0276	0.0304	0.0330
4200 kgf/cm^2	0.0153	0.0184	0.0214	0.0236	0.0256

3.7 MINIMUM REINFORCEMENT RATIO ρ

When the steel reinforcement in a flexural member is only a small amount because the factored moment M_u is low, the beam may perform in service in its uncracked state. The method used for computing flexural strength, however, is based on the assumption that the tension concrete is cracked. Thus, it would be possible for the computed nominal strength M_n for a section having a small amount of reinforcement, based on cracked tension concrete, to be less than the strength M_n (called M_{cr}) of the same section as a plain concrete beam (i.e., having no steel reinforcement). Since a ductile failure mode is desired, the lowest amount of steel permitted should be the amount that would equal the strength of an unreinforced beam. The desired relationship may be developed as follows:

$$\begin{bmatrix} \text{strength of reinforced} \\ \text{concrete beam, } \phi M_n \end{bmatrix} \geq \begin{bmatrix} \text{strength of plain} \\ \text{concrete beam, } M_{cr} \end{bmatrix} \tag{3.7.1}$$

The strength of a plain concrete beam is referred to as the *cracking moment* M_{cr} and is achieved when the extreme fiber in tension reaches the modulus of rupture f_r (see Section 1.8). For normal weight concrete, the ACI Code uses

$$f_r = 7.5\sqrt{f'_c} \tag{3.7.2}$$

3.7 Minimum Reinforcement Ratio ρ :

* When the amount of steel reinforcement in a flexural member is <u>low</u>, it may be possible for the computed nominal strength M_n for a section to be <u>less</u> than the strength M_n (called M_{cr}) of the same section as a plain concrete beam (i.e. having no steel reinforcement).

* This is because the tension concrete is assumed cracked and its tensile strength is neglected.

* Since a ductile mode of failure is desired, the lowest amount of steel permitted should be the amount that would equal the strength of an unreinforced beam.

$$\left\{ \begin{array}{c} \text{strength of reinforced} \\ \text{concrete beam, } \phi M_n \end{array} \right\} \;\geq\; \left(\begin{array}{c} \text{strength of plain} \\ \text{concrete beam, } M_{cr} \end{array} \right)$$

* M_{cr} : cracking moment $\equiv$ flexural strength of a plain concrete beam.

* M_{cr} is achieved when the extreme fiber in tension reaches the modulus of rupture, f_r.

* For normal weight concrete, the ACI Code uses: (Section 9.5.2.3)

$$f_r = 7.5 \sqrt{f_c'} \qquad , \; f_c' \text{ in psi}$$

$$\underset{=}{\text{or}} \qquad f_r = 0.62 \sqrt{f_c'} \qquad , \; f_c' \text{ in MPa}.$$

* Consider a plain (unreinforced) concrete beam (elastic, homogeneous). We have the flexure formula:

$$f = \frac{My}{I}$$

At the maximum moment M_{cr}, we have $f = f_r$:

$$f_r = \frac{M_{cr}\, y_t}{I_g}$$

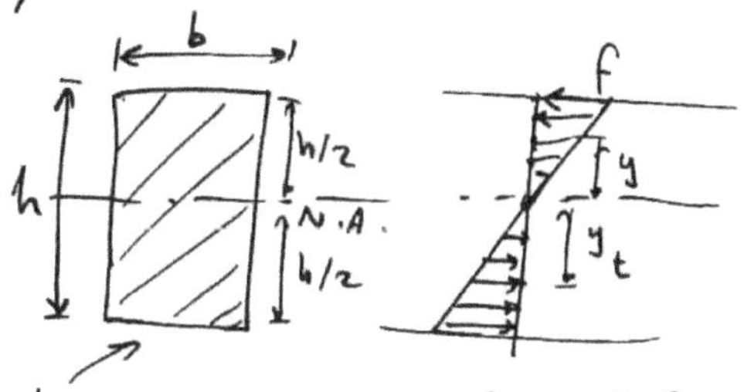

$I_g \equiv$ moment of inertia of the gross concrete section

$y_t \equiv$ distance from the neutral axis to the extreme fiber in tension

$$\therefore \quad M_{cr} = f_r \frac{I_g}{y_t}$$

For a rectangular section, $I_g = \frac{1}{12} b h^3$, $y_t = c_t = \frac{h}{2}$.

$$\therefore \quad M_{cr} = \left(0.62\sqrt{f_c'}\right) \frac{\frac{1}{12} b h^3}{\frac{h}{2}} = \frac{0.62\sqrt{f_c'} \, b h^2}{6} \quad \left.\begin{array}{l}\text{design strength}\\ \text{of a plain}\\ \text{concrete beam}\end{array}\right.$$

In **English** units, $\quad M_{cr} = \frac{7.5\sqrt{f_c'} \, b h^2}{6}$

For a reinforced concrete beam:

$$\phi M_n = \phi \, T \left(d - \frac{a}{2}\right) = \phi \, A_s f_y \left(d - \frac{a}{2}\right)$$

But for minimum steel reinforcement, we must have:

$$\phi M_n \geq M_{cr}$$

$$\phi A_s f_y \left(d - \frac{a}{2}\right) \geq \frac{0.62\sqrt{f_c'} \, b h^2}{6}$$

but $A_s = \rho \, b d$.

For small values of ρ, let $\frac{a}{2} \approx 5\% \, d = 0.05 \, d$.

For flexure, $\phi = 0.90$:

$$0.9 \, (\rho b d) \, f_y \, (d - 0.05 d) \geq \frac{0.62}{6}\sqrt{f_c'} \, b h^2$$

$$\therefore \quad 0.9 \, \rho \, f_y \, d^2 \, (0.95) \geq \frac{0.62}{6}\sqrt{f_c'} \, h^2$$

$$\therefore \quad \rho \geq \frac{0.121\sqrt{f_c'}}{f_y} \left(\frac{h}{d}\right)^2 \quad, \text{ in MPa units.}$$

In English units, $\rho \geq \frac{1.462\sqrt{f_c'}}{f_y} \left(\frac{h}{d}\right)^2$, in psi units.

thus, if $d \approx 0.9 \, h$, i.e. $\frac{h}{d} = 1.1111$

$$\therefore \quad \rho_{min} \geq \frac{(1.462)(1.111)^2 \sqrt{f_c'}}{f_y} = \frac{1.80\sqrt{f_c'}}{f_y},$$

$$f_c', f_n \text{ in psi}$$

$$\rho_{min} \geqslant \frac{0.121 \, (1.111)^2 \, \sqrt{f_c'}}{f_y} = \frac{0.149 \sqrt{f_c'}}{f_y} \, , \quad f_c', f_y \text{ in MPa}.$$

For various values of f_c', ρ_{min} is:

$$f_c' = 3000 \text{ psi} \longrightarrow \rho_{min} = \frac{99}{f_y}$$

$$f_c' = 4000 \text{ psi} \longrightarrow \rho_{min} = \frac{114}{f_y}$$

$$f_c' = 5000 \text{ psi} \longrightarrow \rho_{min} = \frac{128}{f_y}$$

$$f_y \text{ in psi.}$$

$$f_c' = 20 \text{ MPa} \longrightarrow \rho_{min} = \frac{0.666}{f_y}$$

$$f_c' = 30 \text{ MPa} \longrightarrow \rho_{min} = \frac{0.816}{f_y}$$

$$f_c' = 35 \text{ MPa} \longrightarrow \rho_{min} = \frac{0.881}{f_y}$$

$$f_y \text{ in MPa.}$$

ACI Code section 10.5.1 requires $\rho_{min} \geqslant \dfrac{200}{f_y}$, f_y in psi.

$$\rho_{min} \geqslant \frac{1.4}{f_y} \, , \quad f_y \text{ in MPa.}$$

the above ACI equation for ρ_{min} is very conservative.

* the ACI Code is __very__ conservative in this respect for isolated rectangular beams.

* However, it is unconservative in this respect for T-beams (with slab in tension).
   [this will be discussed later in Chapter 9 on T-beams).

---

* the ACI Code (Section 10.5.2) waivers the above requirement on ρ_{min} whenever $\phi M_n \geqslant 1.33 \, M_u$.

* The philosophy is that extra safety is a satisfactory substitute for having a ductile failure mode.

3.8 Design of Rectangular Sections in Bending with Tension Reinforcement Only:

* To design rectangular sections in bending with tension reinforcement only, means to determine the <u>three</u> quantities

$$\underline{\underline{b, d, A_s}} \quad \text{given:} \quad \begin{cases} 1. & M_n = M_u/\phi \\ 2. & f_c' \\ 3. & f_y \end{cases}$$

* No. of unknowns $= 3$
  No. of equations of equilibrium $= 2 \longrightarrow \begin{cases} C = T \\ M_n = C\left(d - \frac{a}{2}\right) \equiv T\left(d - \frac{a}{2}\right) \end{cases}$

* There are <u>many</u> possible solutions.

---

* If the reinforcement ratio ρ is known,

$$C = T$$

$$0.85 f_c' \, a b = A_s f_y = \rho b d f_y$$

$$\therefore \quad a = \rho \left(\frac{f_y}{0.85 f_c'}\right) d$$

$$\therefore \quad M_n = T\left(d - \frac{a}{2}\right) = \rho b d f_y \left[d - \frac{1}{2}\rho\left(\frac{f_y}{0.85 f_c'}\right)d\right]$$

$$= \rho b d^2 f_y \left[1 - \frac{1}{2}\rho\left(\frac{f_y}{0.85 f_c'}\right)\right]$$

Divide by bd^2:

$$\frac{M_n}{bd^2} = \rho f_y \left[1 - \frac{1}{2}\rho\left(\frac{f_y}{0.85 f_c'}\right)\right]$$

let $m = \dfrac{f_y}{0.85 f_c'}$

$$\Longrightarrow \quad \frac{M_n}{bd^2} = \rho f_y \left(1 - \frac{1}{2}\rho m\right)$$

Define $\dfrac{M_n}{bd^2} \equiv R_n$: Coefficient of Resistance.

$$\boxed{R_n = \frac{M_n}{bd^2} = \rho f_y \left(1 - \frac{1}{2}\rho m\right)}$$

See the figure for the relationship between ρ and R_n for various values of f_c', f_y.

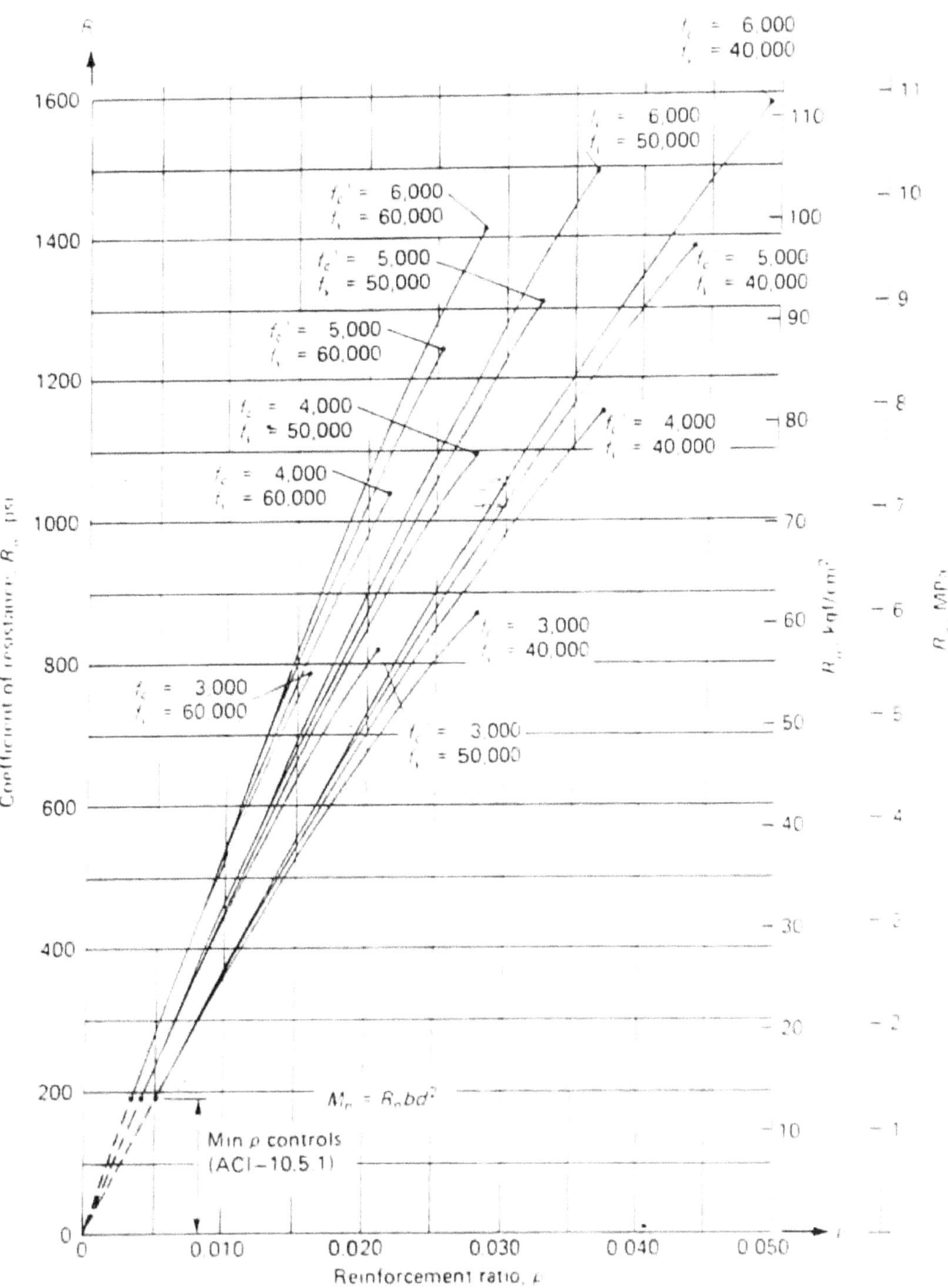

Figure 3.8.1 Strength curves (R_n vs ρ) for singly reinforced rectangular sections. Upper limit of curves is at $0.75\rho_b$.

* In some cases, the values of b and d are given.

Thus R_n is given.

$$\text{thus} \quad R_n = \rho f_y \left(1 - \tfrac{1}{2}\rho m\right) \quad \text{is known.}$$

$$\therefore \quad R_n = \rho f_y - \tfrac{1}{2}\rho^2 f_y m$$

$$\therefore \quad \rho^2 f_y m - 2\rho f_y + 2 R_n = 0$$

$$\therefore \quad \rho = \frac{2 f_y \pm \sqrt{4 f_y^2 - 4(f_y m)(2 R_n)}}{2(f_y m)}$$

$$= \frac{2 f_y}{2 f_y m} \pm \frac{2\sqrt{f_y^2 - 2 m R_n f_y}}{2 f_y m}$$

$$= \frac{1}{m} \pm \frac{1}{m}\sqrt{\frac{f_y^2 - 2 m R_n f_y}{f_y^2}}$$

$$\boxed{\rho = \frac{1}{m}\left(1 \pm \sqrt{1 - \frac{2 m R_n}{f_y}}\right)}$$

Steps in Design:

(1) Assume a value of ρ between the maximum and minimum limits : $\rho_{min} \leq \rho \leq \rho_{max}$

$$\rho_{max} = 0.75\, \rho_b$$

$$\left\{ \begin{array}{l} \text{where} \quad \rho_b = \dfrac{0.85 f_c'}{f_y}\beta_1\left(\dfrac{600}{600 + f_y}\right) \quad , \quad f_y \text{ in } MPa. \\[3mm] \underset{\cong}{or} \quad \rho_b = \dfrac{0.85 f_c'}{f_y}\beta_1\left(\dfrac{87000}{87000 + f_y}\right) , \quad f_y \text{ in } psi. \end{array} \right.$$

$$\rho_{min} = \frac{200}{f_y} \quad , \quad f_y \text{ in } psi$$

$$\rho_{min} = \frac{1.4}{f_y} \quad , \quad f_y \text{ in } MPa.$$

(2) Determine $m = \dfrac{f_y}{0.85 f_c'}$ then $R_n = \rho f_y\left(1 - \tfrac{1}{2}\rho m\right)$.

$$\therefore \text{ Required } bd^2 = \frac{\text{Required } M_n}{R_n}$$

(3) Choose a suitable set of values of b and d so that the provided bd^2 is approximately equal to the required bd^2. [Actually, h is chosen, not d].

14) Compute $R_n = \dfrac{M_n}{bd^2}$ for the selected section, using one of the following methods :

    (a) By the formula (exact) : $\rho = \dfrac{1}{m}\left(1 \pm \sqrt{1 - \dfrac{2\,m\,R_n}{f_y}}\right)$

    (b) By curves (the figure).

    (c) By approximate proportion
        (on the safe side if revised R_n is smaller than original R_n).

$$\rho \approx (\text{original } \rho)\,\dfrac{(\text{revised } R_n)}{(\text{original } R_n)}$$

(5) Compute $A_s = \rho\,bd$ where $\rho \equiv (\text{revised } \rho)$.

(6) Select reinforcement from tables and check the strength of the section to be certain that

$$M_u \le \phi M_n$$

Example 4 :

Determine a set of values of b, d, and A_s that will carry a factored bending moment M_u of 530 kN·m . Use $f_c' = 30$ MPa and $f_y = 300$ MPa.

Solution :

(1) Determine the limits ρ_{min} and ρ_{max} :

$$\rho_b = \dfrac{0.85\,f_c'}{f_y}\,\beta_1 \left(\dfrac{600}{600 + f_y}\right) \quad , \quad f_y \text{ in MPa}$$

where $\beta_1 = 0.85$ when $f_c' = 30$ MPa .

$$\therefore\quad \rho_b = \dfrac{0.85\,(30)}{(300)}\,(0.85)\left(\dfrac{600}{600 + 300}\right) = 0.0482 = 4.82\% .$$

$$\therefore\quad \rho_{max} = 0.75\,\rho_b = 0.75\,(0.0482) = 0.03615 = 3.615\% .$$

$$\rho_{min} = \dfrac{1.4}{f_y} = \dfrac{1.4}{300} = 0.00467 = 0.467\% \approx 0.5\% .$$

$$\therefore\quad 0.005 \le \rho \le 0.03615$$

Take $\rho = 0.03 = 3\%$.

(2) $\quad m = \dfrac{f_y}{0.85 f_c'} = \dfrac{300}{0.85(30)} = 11.76$.

$$R_n = \rho f_y \left(1 - \tfrac{1}{2}\rho m\right) = (0.03)(300)\left[1 - \tfrac{1}{2}(0.03)(11.76)\right]$$

$$= 7.41 \; MPa .$$

$$M_u \le \phi M_n .$$

$$\therefore \quad M_n \geqslant \dfrac{M_u}{\phi} = \dfrac{530}{0.9} = 589 \; kN\cdot m \qquad \left(\phi = 0.9 \text{ for flexure}\right).$$

$$R_n = \dfrac{M_n}{bd^2} \implies bd^2 = \dfrac{M_n}{R_n} \geqslant \dfrac{589 \times 1000 \times 1000}{7.41 \times 10^6 \times 10^{-6}}$$

$$= 79.487 \times 10^6 \; mm^3 .$$

(3) Now select b and d :

$$\text{Take} \quad b = 350 \; mm \implies bd^2 = 79.487 \times 10^6$$

$$\therefore d^2 = \dfrac{79.487 \times 10^6}{350} \implies d = 476.6 \; mm .$$

$$\therefore \text{Take } d = 500 \; mm .$$

(d is usually $60-65$ mm less than h, if one layer of steel is used).

(4) $\quad R_n = \dfrac{M_n}{bd^2} = \dfrac{589 \times 10^3}{\left(\frac{350}{1000}\right)\left(\frac{500}{1000}\right)^2} = 6.73 \times 10^6 \; N/m^2$

$$= 6.73 \; MPa .$$

$$\rho = \dfrac{1}{m}\left(1 \pm \sqrt{1 - \dfrac{2m R_n}{f_y}}\right) = \dfrac{1}{11.76}\left(1 \pm \sqrt{1 - \dfrac{2(11.76)(6.73)}{300}}\right)$$

$$= 0.0266 \quad = 2.66 \% .$$

(5) $A_s = \rho bd = (0.0266)(350)(500) = 4655 \; mm^2 = 46.55 \; cm^2 .$

(6) Select reinforcement from the table :

$$\text{Take } 2\text{-}32\,mm \text{ and } 3\text{-}36\,mm \text{ bars (two layers).}$$

$$A_s = 16.08 + 30.54 = 46.62 > 46.55 \; cm^2 \quad \longleftarrow$$

<u>Check the design :</u>

$$T = A_s f_y = 46.62 \times (10^{-2})^2 \times 300 \times 1000 = 1399 \; kN .$$

$$C = T = 0.85 f_c' a b$$

$$\therefore a = \frac{T}{0.85 f_c' b} = \frac{1399 \times 10^3}{0.85 (30 \times 10^6 \times (10^{-3})^2) \times 350} = 157 \text{ mm.}$$

$$\therefore M_n = T \left(d - \frac{a}{2} \right) = 1399 \left(\frac{500}{1000} - \frac{157}{2(1000)} \right) = 590 \text{ kN·m}$$

Check that $\qquad M_u \leq \phi M_n$

$$530 \leq 0.9 (590) = 531 \text{ kN·m} \quad \smile$$

Final decision and design sketch:

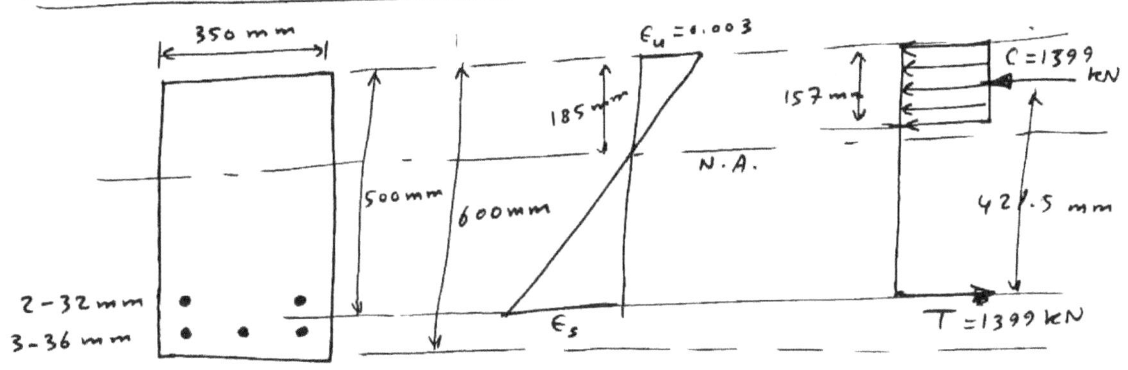

Neutral axis: $\qquad a = \beta_1 x \implies \qquad 157 \text{ mm} = 0.85 x \implies x = 185 \text{ mm.}$

$\therefore$ Use 3-36 mm and 2-32 mm bars in the 350 by 500 mm beam, shown in the figure.

Note: * $\rho = 2.66\%$ is a high percentage of reinforcement. Thus the section 350 mm $\times$ 500 mm is smaller than is usual.

* For practical bar placement in one layer, and the need for larger cross-section to reduce deflection, ρ does not usually exceed $0.5 \rho_{max}$.

* Thus deflection must be checked later.

Chapter 3
Strength of Rectangular Sections in Bending
Part II

3.9 Practical Selection for Beam Sizes, Bar Sizes, and Bar Placement:

* Choice of steel reinforcement ratio ρ depends on the limitation on the deflection of the beam.

* Deflection problems are rarely encountered with beams having ρ not more than one-half the maximum permissible value $\left[\ i.e.\ \frac{1}{2}(0.75\rho_b) = 0.375\rho_b\right]$.

* For the placement of bars within the beam width, ACI—Code (Section 7.6.1) specifies the clearance needed between bars to permit proper concrete placement around them.

* This clearance for bars in a layer is 1 in. (25.4 mm) or the nominal diameter of the bar, whichever is greater.

* When two or more layers are required, the minimum clearance between layers is 1 in. (25.4 mm). [ACI—Code, Section 7.6.2].

* The bars in different layers must be located directly above one another and not staggered [ACI Code, Section 7.6.2].

* The following are some guidelines in making choices for beam sizes, bar sizes, and bar placement. Note: these are not ACI Code requirements.

① For beam sizes :

  1. Use whole numbers (mm) for overall dimensions.

  2. An __economical__ rectangular beam proportion is one in which the overall depth-to-width ratio, h/b, is between about 1.5 to 2.0 .

  3. For T-shaped beams, typically the flange thickness represents about 20% of overall depth.
  (See Chapter 9).

② For reinforcing bars:

1. Maintain bar symmetry about the centroidal axis which lies at right angles to the bending axis.

2. Use at least _two_ bars wherever flexural reinforcement is required.

3. Use #11 bars ($\approx$ 35 mm diameter) and smaller for usual sized beams.

4. Use no more than _two_ bar sizes and no more than two standard sizes apart for steel in one face at a given location in the span (i.e. #7 and #9 bars may be acceptable, but #9 and #4 bars would not). [22 mm and 28 mm are o.k., but 12 mm and 28 mm are _not_ o.k.].

5. Place bars in one layer if practicable. Try to select bar sizes so that no less than two and no more than five or six bars are put in one layer.

6. Follow the requirements of ACI – 7.6.1 and ACI 7.6.2 for clear distance between bars and between layers, and arrangement between layers.

7. When different sizes of bars are used in several layers at a location, place the largest bars in the layer nearest the face of the beam.

Example 5:

Select an economical rectangular beam size and select bars using the ACI strength method. The beam is a simply-supported span of 12 m and it is to carry a live load of 19 kN/m and a dead load of 11.5 kN/m (_not_ including beam weight). Without actually checking deflections, use a reinforcement ratio ρ such that excessive deflection is unlikely. Use $f_c' = 30$ MPa, and $f_y = 400$ MPa.

Table 3.9.1 Total Areas for Various Numbers of Reinforcing Bars

BAR SIZE	NOMINAL DIAMETER (in.)	WEIGHT (lb/ft)	NUMBER OF BARS									
			1	2	3	4	5	6	7	8	9	10
#3	0.375	0.376	0.11	0.22	0.33	0.44	0.55	0.66	0.77	0.88	0.99	1.10
#4	0.500	0.668	0.20	0.40	0.60	0.80	1.00	1.20	1.40	1.60	1.80	2.00
#5	0.625	1.043	0.31	0.62	0.93	1.24	1.55	1.86	2.17	2.48	2.79	3.10
#6	0.750	1.502	0.44	0.88	1.32	1.76	2.20	2.64	3.08	3.52	3.96	4.40
#7	0.875	2.044	0.60	1.20	1.80	2.40	3.00	3.60	4.20	4.80	5.40	6.00
#8	1.000	2.670	0.79	1.58	2.37	3.16	3.95	4.74	5.53	6.32	7.11	7.90
#9	1.128	3.400	1.00	2.00	3.00	4.00	5.00	6.00	7.00	8.00	9.00	10.00
#10	1.270	4.303	1.27	2.54	3.81	5.08	6.35	7.62	8.89	10.16	11.43	12.70
#11	1.410	5.313	1.56	3.12	4.68	6.24	7.80	9.36	10.92	12.48	14.04	15.60
#14^a	1.693	7.65	2.25	4.50	6.75	9.00	11.25	13.50	15.75	18.00	20.25	22.50
#18^a	2.257	13.60	4.00	8.00	12.00	16.00	20.00	24.00	28.00	32.00	36.00	40.00

^{a} #14 and #18 bars are used primarily as column reinforcement and are rarely used in beams.

have presented a sensitivity study to show how the variables in simply supported beam design affect the economics of the design.

For the selection of an actual number of bars to meet a total steel area requirement. it is desirable to tabulate the combined area of several bars at a time. Table 3.9.1 gives bar areas for up to 10 bars of different sizes. Note that once the area of one bar, say of #8, is set at 0.79 sq in., the area of 10 bars becomes 7.90, not 7.85; this practice has been carried on by tradition.

For the placement of bars within the beam width, ACI-7.6.1 specifies the clearance needed between bars to permit proper concrete placement around them. This clearance for bars in a layer is 1 in. or the nominal diameter of the bar, whichever is greater. When two or more layers of bars are required, the minimum clearance between layers is 1 in. (ACI-7.6.2). The bars in different layers must be located directly above one another and not staggered (ACI-7.6.2). Table 3.9.2 gives minimum beam widths for various numbers of equal-sized bars, computed in the manner described above.

Since the tensile force in reinforcing bars requires an interaction of the bars with the surrounding concrete, larger than minimum spacing and clear cover of bars will make the concrete confine the bars more efficiently. This benefit is explicitly recognized for the first time in the 1989 ACI Code. As will be discussed in Chapter 6 on development of the tension force in the bars, the use of at least $2d_b$ clear lateral spacing of bars (a category of improved performance) or $3d_b$ clear lateral spacing of bars (the best category of performance) will accrue benefit when determining the lengths of reinforcing bars to be used. Since the selection of cross-section and bars is all that has been treated thus far in this chapter, suffice it to say that shorter length bars can be used when $2d_b$ clear lateral spacing can be used (minimum width in Table 3.9.3), and when $3d_b$ clear lateral spacing can be used (minimum width in Table 3.9.4) the minimum lengths of reinforcement will be required.

Table 3.9.2 Minimum Beam Width (Inches) According to the ACI Code[a]

SIZE OF BARS	NUMBER OF BARS IN SINGLE LAYER OF REINFORCEMENT							ADD FOR EACH ADDED BAR
	2	3	4	5	6	7	8	
#4	6.8	8.3	9.8	11.3	12.8	14.3	15.8	1.50
#5	6.9	8.5	10.2	11.8	13.4	15.0	16.7	1.63
#6	7.0	8.8	10.5	12.3	14.0	15.8	17.5	1.75
#7	7.2	9.0	10.9	12.8	14.7	16.5	18.4	1.88
#8	7.3	9.3	11.3	13.3	15.3	17.3	19.3	2.00
#9	7.6	9.8	12.2	14.3	16.6	18.8	21.1	2.26
#10	7.8	10.4	12.9	15.5	18.0	20.5	23.1	2.54
#11	8.1	10.9	13.8	16.6	19.4	22.2	25.0	2.82
#14	8.9	12.3	15.7	19.1	22.5	25.9	29.3	3.40
#18	10.6	15.1	19.6	24.1	28.6	33.1	37.6	4.51

Table shows minimum beam widths when #3 stirrups are used.

For additional bars, add dimension in last column for each added bar.

For bars of different size, determine from table the beam width for smaller size bars and then add last column figure for each larger bar used.

[a] Assumes maximum aggregate size does not exceed three-fourths of the clear space between bars (ACI–3.3.2). Table computation procedure is in agreement with the ACI Code interpretation of ACI Committee 340, as used in the *Strength Design Handbook* [2.21].

$A = 1\frac{1}{2}$-in. clear cover to stirrup

$B = \frac{3}{8}$-in. stirrup bar diameter

$C =$ For #11 and smaller bars, use twice the diameter of #3 stirrups (i.e., $C = 0.75$ in.). For #14 and #18 bars, use $C = 0.5d_b$

$D =$ clear distance between bars $= d_b$ or 1 in., whichever is greater (where d_b is the diameter of the larger adjacent longitudinal bar)

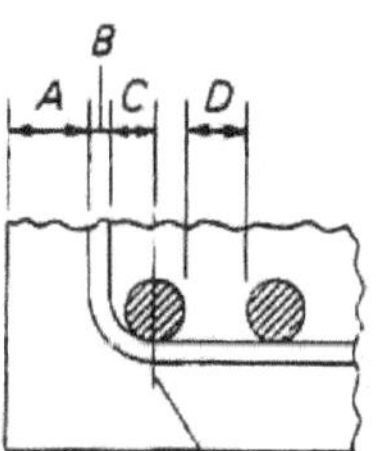

Diameter of corner bar is assumed to be located to intersect the horizontal tangent to stirrup bend

Table 3.9.3 Minimum Beam Width (inches) To Satisfy
2 Bar Diameters Clear Spacing

BAR SIZE	NUMBER OF BARS IN SINGLE LAYER						
	2	3	4	5	6	7	8
#4	6.8	8.3	9.8	11.3	12.8	14.3	15.8
#5	7.1	9.0	10.9	12.8	14.6	16.5	18.4
#6	7.5	9.8	12.0	14.3	16.5	18.8	21.0
#7	7.9	10.5	13.1	15.8	18.4	21.0	23.6
#8	8.3	11.3	14.3	17.3	20.3	23.3	26.3
#9	8.6	12.0	15.4	18.8	22.2	25.6	28.9
#10	9.1	12.9	16.7	20.5	24.3	28.1	31.9
#11	9.5	13.7	17.9	22.2	26.4	30.6	34.9
#14	12.2	15.9	20.9	26.0	31.1	36.2	41.2
#18	15.0	19.8	26.6	33.3	40.1	46.9	53.7

Table Assumptions:

a. Side cover 1.5 in. each side.
b. #3 stirrups for bars #11 and smaller.
c. #4 stirrups for bars #14 and #18.
d. Since stirrups are bent around 4 stirrup bar diameters, the distance from centroid of bar nearest side face of beam to inside face of #3 stirrup is taken as 0.75 in. for bars #11 and smaller, and equal to the bar radius for #14 and #18 bars.

Table 3.9.4 Minimum Beam Width (inches) To Satisfy
3 Bar Diameters Clear Spacing

BAR SIZE	NUMBER OF BARS IN SINGLE LAYER						
	2	3	4	5	6	7	8
#4	7.3	9.3	11.3	13.3	15.3	17.3	19.3
#5	7.8	10.3	12.8	15.3	17.8	20.3	22.8
#6	8.3	11.3	14.3	17.3	20.3	23.3	26.3
#7	8.8	12.3	15.8	19.3	22.8	26.3	29.8
#8	9.3	13.3	17.3	21.3	25.3	29.3	33.3
#9	9.8	14.3	18.8	23.3	27.8	32.3	36.8
#10	10.3	15.4	20.5	25.6	30.7	35.7	-40.8
#11	10.9	16.5	22.2	27.8	33.5	39.1	44.7
#14	12.5	19.2	26.0	32.8	39.6	46.3	53.1
#18	15.3	24.3	33.3	42.4	51.4	60.4	69.5

* Table Assumptions:

a. Side cover 1.5 in. each side.
b. #3 stirrups for bars #11 and smaller.
c. #4 stirrups for bars #14 and #18.
d. Since stirrups are bent around 4 stirrup bar diameters, the distance from centroid of bar nearest side face of beam to inside-face of #3 stirrup is taken as 0.75 in. for bars #11 and smaller, and equal to the bar radius for #14 and #18 bars.

Weight, Perimeter and Area of Metric Steel Bars

Bar diameter,	Weight per meter,	Peri-meter,	Area in mm² for various numbers of bars											
mm	kg	mm	1	2	3	4	5	6	7	8	9	10	11	12
5	0.15	15.7	20	39	59	78	98	118	137	157	177	196	216	235
6	0.22	18.8	28	56	85	113	141	170	198	226	254	283	311	340
8	0.39	25.1	50	100	151	201	251	301	352	402	452	503	553	604
10	0.62	31.4	79	157	236	314	393	471	550	628	707	785	864	942
12	0.89	37.7	113	226	339	452	565	678	791	905	1018	1131	1244	1357
14	1.21	44.0	154	308	462	616	770	724	1078	1232	1386	1539	1694	1848
16	1.58	50.3	201	402	603	804	1005	1206	1407	1608	1809	2011	2212	2413
18	2.00	56.5	254	509	763	1018	1272	1527	1781	2036	2290	2545	2799	3054
20	2.47	62.8	314	628	942	1257	1571	1884	2199	2514	2828	3142	3456	3770
22	2.98	69.1	380	760	1140	1521	1901	2281	2661	3041	3421	3801	4181	4562
25	3.85	78.5	491	982	1473	1963	2454	2945	3436	3927	4418	4909	5400	5891
28	4.83	88.0	616	1231	1847	2463	3079	3694	4310	4926	5542	6158	6773	7389
30	5.54	94.2	707	1414	2120	2827	3534	4241	4948	5654	6361	7068	7775	8482
32	6.31	100.5	804	1608	2413	3217	4021	4826	5630	6434	7238	8042	8847	9651
36	7.99	113.1	1018	2036	3054	4072	5090	6107	7126	8143	9161	10179	11197	12215
40	9.86	125.7	1257	2513	3770	5026	6283	7540	8796	10053	11309	12566	13823	15080
45	12.49	141.4	1590	3181	4771	6362	7952	9542	11133	12732	14314	15904	17494	19085
50	15.41	157.1	1964	3927	5891	7854	9818	11781	13745	15708	17672	19635	21599	23562

<u>Solution</u> :

(a) Decide on a reinforcement ratio ρ :

Choose $\rho = \frac{1}{2}\rho_{max.} = \frac{1}{2}(0.75\rho_b) = 0.375\rho_b$.

[Either calculate ρ_b or obtain it from tables] .

For $f'_c = 30\ MPa \implies \beta_1 = 0.85$.

$\therefore\quad \rho_b = \dfrac{0.85\,f'_c}{f_y}\,\beta_1\left(\dfrac{600}{600+f_y}\right) = \dfrac{0.85\,(30)}{400}(0.85)\left(\dfrac{600}{600+400}\right)$

$\qquad\qquad = 0.0325$

$\therefore\quad \rho_{max.} = 0.75\,(0.0325) = 0.0244$

$\therefore\quad$ Use $\rho = \dfrac{0.0244}{2} = 0.012 = 1.2\%$.

(b) Determine R_n :

$\qquad m = \dfrac{f_y}{0.85\,f'_c} = \dfrac{400}{0.85\,(30)} = 15.69$.

$\qquad R_n = \rho f_y\left(1-\tfrac{1}{2}\rho m\right) = 0.012\,(400)\left[1-\tfrac{1}{2}(0.012)(15.69)\right]$

$\qquad\qquad\qquad\qquad\qquad = 4.35\ MPa$.

(c) Determine factored moments:

Estimate the weight of the beam.

Assume $b = 450\ mm$, $d = 500\ mm$.

$\qquad$ Use $25\ kN/m^3$ for reinforced concrete.

$\qquad w = 25 \times \dfrac{450}{1000} \times \dfrac{500}{1000} \times 1 \approx 5.625\ kN/m$.

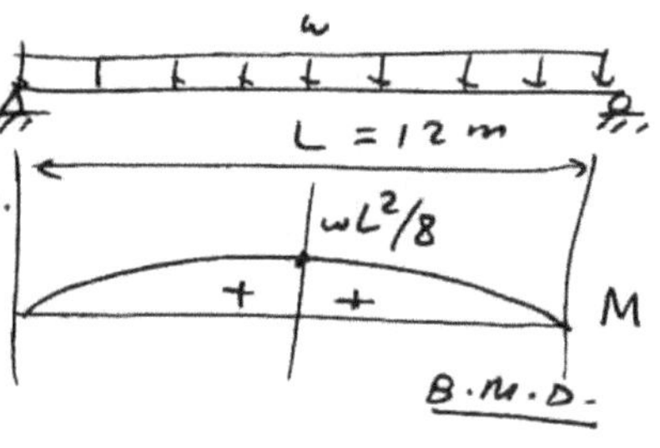

$\therefore\ w_d = 5.625 + 11.5 = 17.125\ kN/m$.

$\underline{w_\ell = 19\ kN/m}$.

$M_d = \dfrac{w_d L^2}{8} = \dfrac{(17.125)(12)^2}{8} = 308.25\ kN\cdot m$.

$M_\ell = \dfrac{w_\ell L^2}{8} = \dfrac{(19)(12)^2}{8} = 342\ kN\cdot m$.

$\therefore\ M_u = 1.4\,M_d + 1.7\,M_\ell$

$\qquad\quad = 1.4\,(308.25) + 1.7\,(342)$

$\therefore\ M_u = 1013\ kN\cdot m$.

Required $M_n \geqslant \dfrac{M_u}{\Phi} = \dfrac{1013}{0.9} = 1126$ kN·m

(d) Required $bd^2 = \dfrac{M_n}{R_n} \geqslant \dfrac{1126}{4.35 \times 1000} = 0.25885$ m³

$$= 258.85 \times 10^6 \text{ mm}^3$$

$$\therefore \quad d \geqslant \sqrt{\dfrac{258.85 \times 10^6}{b}}$$

(e) Select width b and determine required depth d.
Make a table as follows:

Chosen b	Required d	Ratio d/b
300 mm	929 mm	3.1
350 mm	860 mm	2.5
400 mm	804 mm	2.0
450 mm	758 mm	1.68
500 mm	720 mm	1.44

Choose the dimensions that will give the <u>most economical</u>
cross-section, i.e. d/b between 1.5 and 2.0.
(according to the guidelines). [Actually h/b]!

$\therefore$ Take $b = 450$ mm , $d = 758$ mm.

Now, determine the overall depth, h.

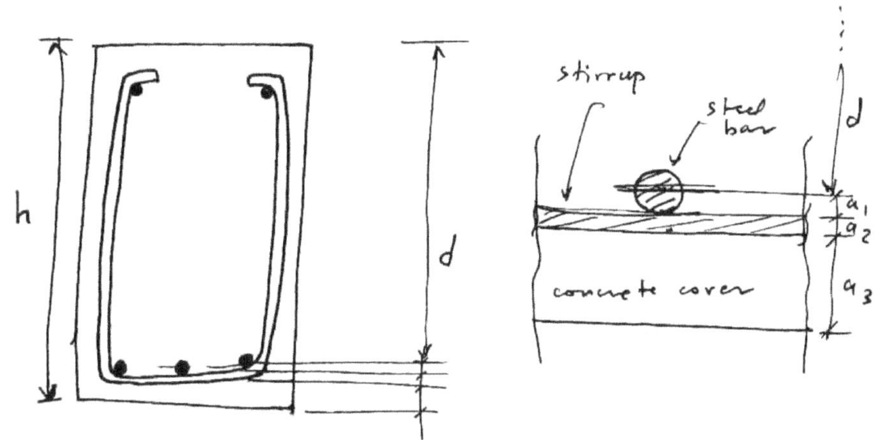

a_1 : bar radius
a_2 : stirrup (usually 9.5 mm – 12.5 mm, diameter).
a_3 : clear cover (40 mm, minimum) (<u>not</u> exposed).

$h = d + a_1 + a_2 + a_3$

$\quad = 758 + 40 + 9.5 + 12.5$ $\qquad$ (take $a_3 = 12.5$ mm)
bar radius

$\quad = 820$ mm.

(Usually $\;h = d + 60$ to 65 mm).

See ACI Code section 7.7.1 bar cover

The overall depth should be in whole mm.

$\quad$ Take $\boxed{h = 850 \text{ mm}}$.

(You can use a somewhat less or somewhat more value than the computed h).

(f) Check the weight and revise M_u.

$$w = 25 \times \frac{450}{1000} \times \frac{850}{1000} = 9.5625 \text{ kN/m}.$$

$$\therefore \; w_d = 9.5625 + 11.5 = 21.0625 \text{ kN/m}.$$

$$\therefore \; M_d = \frac{w_d L^2}{8} = \frac{(21.0625)(12)^2}{8} = 379 \text{ kN·m}.$$

$$\therefore \; M_u = 1.4\,M_d + 1.7\,M_\ell = 1.4(379) + 1.7(342)$$
$$= 1112 \text{ kN·m}. \quad (\underline{revised}).$$

Revised required $M_n = \dfrac{M_u}{\phi} = \dfrac{1112}{0.9} = 1236 \text{ kN·m}.$

Compute d from h:

$\quad d = h - 65$ mm $\qquad$ (for one-layer of bars)

$\quad\quad = 850 - 65$

$\quad\quad = 785$ mm.

Required $R_n = \dfrac{\text{Required } M_n}{b d^2} = \dfrac{1236}{\left(\frac{450}{1000}\right)\left(\frac{785}{1000}\right)^2} = 4457 \text{ kN/m}^2$

$$= 4.46 \text{ MPa}.$$

$$\rho = \frac{1}{m}\left(1 - \sqrt{1 - \frac{2m R_n}{f_y}}\right) \qquad (\text{or use the curve}).$$

$$= \frac{1}{15.69}\left(1 - \sqrt{1 - \frac{2(15.69)(4.46 \times 1000)}{400 \times 1000}}\right) = 0.0123 = 1.23\%.$$

$$\rho = \frac{A_s}{bd} \implies A_s = \rho bd = (0.0123)(450)(785) = \underline{4345 \ mm^2} \quad \boxed{58}$$

Make a table of possible alternatives. (Use page $\boxed{54d}$).

Bars	$A_s \ (mm^2)$	Remarks
$10 - \phi 25$	4909	too high, too many bars
$8 - \phi 28$	4926	too high, too many bars
$7 - \phi 30$	4948	too high, too many bars
$7 - \phi 28$	4310	too low, too many bars
$6 - \phi 32$	4826	too high (bars may not fit in one layer).
$4 - \phi 32$ $2 - \phi 28$	(4448)	o.k.
$4 - \phi 32$ $2 - \phi 30$	4631	too high
$3 - \phi 32$ $3 - \phi 28$	4260	too low, not symmetric
$3 - \phi 32$ $3 - \phi 30$	4533	not symmetric

* Check whether $4 - \phi 32$ and $2 - \phi 28$ bars will fit into a 450-mm width in <u>one</u> layer.

* Use 10-mm stirrups.

"Approximate" clear spacing between bars

$$= \frac{450 - 2(40) - 2(10) - 4(32) - 2(28)}{5} = 33.2 \ mm > 32 \ mm.$$
$$\underline{o.k.} \ \smile$$

Note: the clearance for bars in a layer is 25 mm or the nominal diameter of the bars, whichever is greater.

Note: You can use table 3.9.2 page 54b for <u>minimum</u> beam width.

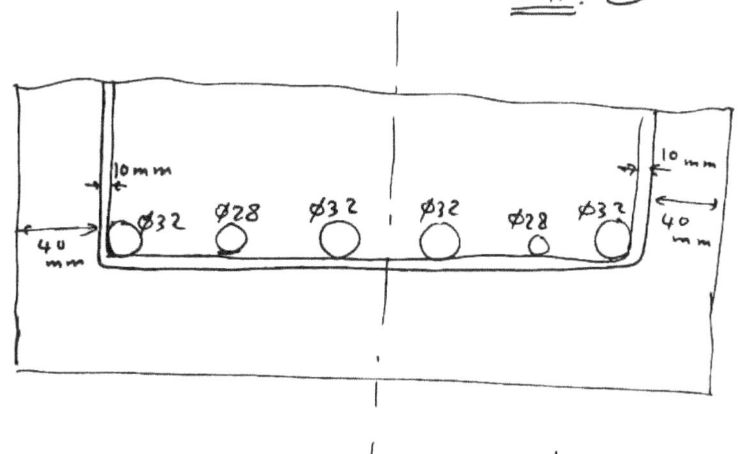

(9) Check strength and provide design sketch:

Use computed $d = 785$ mm and $A_s = 4448$ mm².

$$C = T$$

$$0.85 f_c' a b = f_y A_s \implies 0.85(30)a(450) = (400)(4448)$$

$$\implies a = 155 \text{ mm}.$$

$$M_n = T\left(d - \frac{a}{2}\right) = f_y A_s \left(d - \frac{a}{2}\right)$$

$$= (400 \times 1000)(4448 \times 10^{-6})\left(\frac{785}{1000} - \frac{1}{2} \cdot \frac{155}{1000}\right)$$

$$= 1259 \text{ kN·m}.$$

Design strength $\phi M_n = (0.9)(1259) = 1133$ kN·m.

Required Strength $M_u = 1112$ kN·m $< \phi M_n = 1133$ kN·m.

> Use 450×850 mm beam with $4-\phi32$ and $2-\phi28$ bars as shown in the figure.

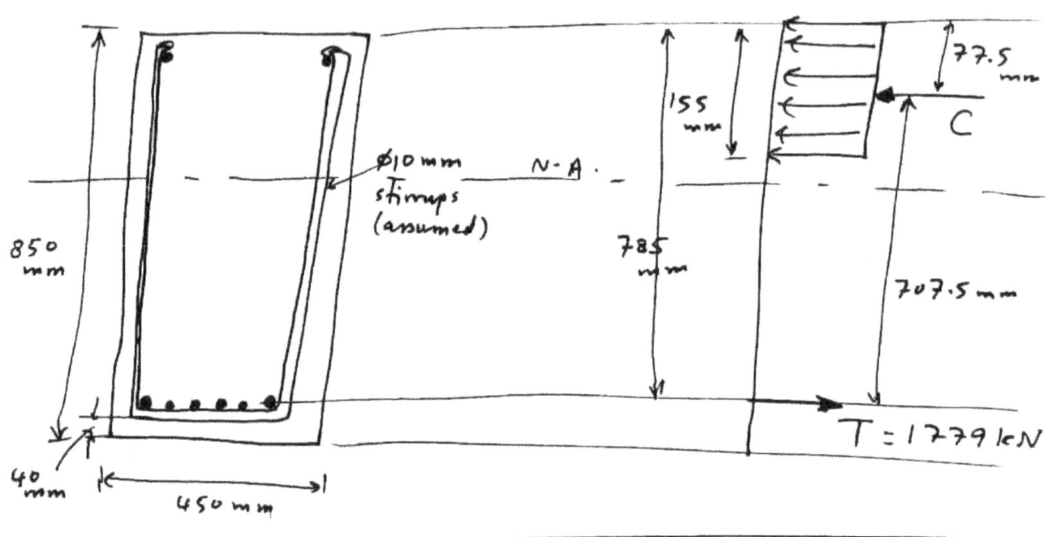

Example 6:

Design a floor slab to carry a uniform live load of 17 kN/m² on a simply-supported span of 7 m, given $f_c' = 20$ MPa, and $f_y = 420$ MPa. Use the ACI Code. Keep the reinforcement ratio at about 30% of the amount in the balanced strain condition so that deflection is unlikely to be excessive.

Solution.

* Note: Take a 1 m-wide typical strip for calculations instead of taking the entire slab width.

* Note: this is known as a one-way slab that acts as a wide beam.

(a) Estimate the dead load of the slab:

take $h = 400$ mm.

$$w_d = 25 \times \frac{400}{1000} \times 1 \times 1 = 10 \text{ kN/m (per meter width)}.$$

$$w_\ell = 17 \times 1 = 17 \text{ kN/m. (per meter width)}.$$

$$w_u = 1.4 w_d + 1.7 w_\ell = 1.4(10) + 1.7(17) = 42.9 \text{ kN/m} \quad \text{(per meter width of slab)}.$$

$$M_u = \frac{w_u L^2}{8} = \frac{(42.9)(7)^2}{8} = 263 \text{ kN·m} \quad \text{(per meter of width)}.$$

$$\therefore \text{Required } M_n = \frac{M_u}{\phi} = \frac{263}{0.9} = 292 \text{ kN·m}.$$

(b) For $f_c' = 20$ MPa, we have $\beta_1 = 0.85$.

$$\rho_b = \frac{0.85 f_c'}{f_y} \beta_1 \left(\frac{600}{600 + f_y} \right) = \frac{0.85(20)}{420} (0.85) \left(\frac{600}{600 + 420} \right)$$

$$= 0.0202 = 2.02 \%.$$

Take $\rho = 30\% \, \rho_b = 0.3 \rho_b = 0.3(0.0202) = 0.00607$
$$= 0.607 \%.$$

$$m = \frac{f_y}{0.85 f_c'} = \frac{420}{0.85(20)} = 24.71.$$

$$R_n = \rho f_y \left(1 - \frac{1}{2} \rho m \right) = (0.00607)(420 \times 1000) \left[1 - \frac{1}{2}(0.00607)(24.71) \right]$$

$$= 2358 \text{ kN/m}^2$$

$$= 2.358 \text{ MPa}.$$

(c) Required $bd^2 = \frac{M_n}{R_n} \geqslant \frac{292}{2358} = 0.1238 \text{ m}^3$
$$= 123.82 \times 10^6 \text{ mm}^3.$$

For a slab, $b = 1 \text{ m} \equiv 1000 \text{ mm}$.

$$\therefore \quad d \geqslant \sqrt{\frac{123.82 \times 10^6}{1000}} = 352 \text{ mm}.$$

Note: the required clear cover for a slab = 20 mm

(**not** exposed, ACI Code, Section 7.7.1)

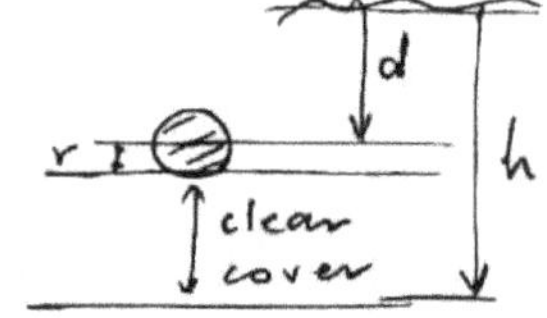

$h = d + $ bar radius $ + $ clear cover

$\qquad = 352 + 12 + 20$

$\qquad = 384 \text{ mm}.$

Take $\quad h = 400 \text{ mm}.$

* Our dead load estimation is correct.

$\Rightarrow$ No need to revise.

(d) Actual $d = h - 20 - 12 = 400 - 20 - 12 = 368 \text{ mm}.$

Required $R_n = \dfrac{M_n}{bd^2} = \dfrac{292}{(1000)(368)^2 \times 10^{-9}} = 2156 \text{ kN/m}^2$

$\qquad\qquad\qquad\qquad\qquad\qquad\qquad = 2.156 \text{ MPa}.$

$$\Rightarrow \rho = \frac{1}{m}\left(1 - \sqrt{1 - \frac{2mR_n}{f_y}}\right) = \frac{1}{24.71}\left(1 - \sqrt{1 - \frac{2(24.71)(2.156)}{420}}\right)$$

$$= 0.00551 \quad = 0.551\%.$$

$A_s = \rho bd = (0.00551)(1000)(368) = 2027 \text{ mm}^2.$

Use the table on page **61a** to select reinforcement.

Use $\phi 22 @ 180 \text{ mm}$ spacing $\Rightarrow A_s = 2112 \text{ mm}^2.$

Note: For slabs, bars are **not** selected by picking a total number; instead, they are selected on the basis of an average area provided per meter of width.

one $\phi 22 \text{ mm} \Rightarrow$ area $= \dfrac{\pi (22)^2}{4} = 380 \text{ mm}^2$
(for 1000 mm spacing)

For 180 mm spacing, $A_s = 380\left(\dfrac{1000}{180}\right) = 2111 \text{ mm}^2.$

Areas of Steel Bars for Various Spacings

Spacing mm	Area in mm² per meter width of slab for bar diameter								
	5 mm	6 mm	8 mm	10 mm	12 mm	16 mm	20 mm	22 mm	25 mm
60	326	472	838	1308	1884	3352	5234	6334	8200
65	302	434	774	1208	1740	3094	4832	5848	7580
70	280	404	718	1122	1616	2872	4488	5430	7020
75	262	378	670	1048	1508	2682	4188	5068	6550
80	246	354	628	982	1414	2514	3928	4752	6150
85	232	332	592	924	1330	2366	3696	4472	5790
90	218	314	558	872	1256	2234	3492	4224	5470
95	206	298	530	826	1190	2116	3308	4002	5160
100	196	283	503	785	1131	2011	3142	3801	4908
105	187	269	479	748	1077	1915	2991	3620	4690
110	178	257	457	714	1028	1828	2855	3455	4470
115	171	246	437	683	984	1749	2731	3305	4275
120	163	236	419	654	942	1676	2617	3167	4100
125	157	226	402	628	905	1609	2513	3041	3940
130	151	217	387	604	870	1547	2416	2924	3790
135	145	209	372	582	838	1490	2327	2816	3650
140	140	202	359	561	808	1436	2244	2715	3510
145	135	195	347	542	780	1387	2166	2621	3400
150	131	189	335	524	754	1341	2094	2534	3275
155	127	182	324	507	730	1297	2027	2452	3175
160	123	177	314	491	707	1257	1964	2376	3075
165	119	171	305	476	685	1219	1904	2304	2980
170	116	166	296	462	665	1183	1848	2236	2895
175	112	162	287	449	646	1149	1795	2172	2810
180	109	157	279	436	628	1117	1746	2112	2735
185	106	153	272	425	611	1087	1694	2055	2655
190	103	149	265	413	595	1058	1654	2001	2580
195	101	145	258	403	580	1031	1611	1949	2520
200	98	141	251	393	565	1005	1571	1901	2454

Table 3.9.6 Average Area per Foot of Width Provided by Various Bar Spacings

BAR SIZE NUMBER	NOMINAL DIAMETER (in.)	SPACING OF BARS IN INCHES													
		2	$2\frac{1}{2}$	3	$3\frac{1}{2}$	4	$4\frac{1}{2}$	5	$5\frac{1}{2}$	6	7	8	9	10	12
3	0.375	0.66	0.53	0.44	0.38	0.33	0.29	0.26	0.24	0.22	0.19	0.17	0.15	0.13	0.11
4	0.500	1.18	0.94	0.78	0.67	0.59	0.52	0.47	0.43	0.39	0.34	0.29	0.26	0.24	0.20
5	0.625	1.84	1.47	1.23	1.05	0.92	0.82	0.74	0.67	0.61	0.53	0.46	0.41	0.37	0.31
6	0.750	2.65	2.12	1.77	1.51	1.32	1.18	1.06	0.96	0.88	0.76	0.66	0.59	0.53	0.44
7	0.875	3.61	2.88	2.40	2.06	1.80	1.60	1.44	1.31	1.20	1.03	0.90	0.80	0.72	0.60
8	1.000		3.77	3.14	2.69	2.36	2.09	1.88	1.71	1.57	1.35	1.18	1.05	0.94	0.78
9	1.128		4.80	4.00	3.43	3.00	2.67	2.40	2.18	2.00	1.71	1.50	1.33	1.20	1.00
10	1.270			5.06	4.34	3.80	3.37	3.04	2.76	2.53	2.17	1.89	1.69	1.52	1.27
11	1.410			6.25	5.36	4.69	4.17	3.75	3.41	3.12	2.68	2.34	2.08	1.87	1.56

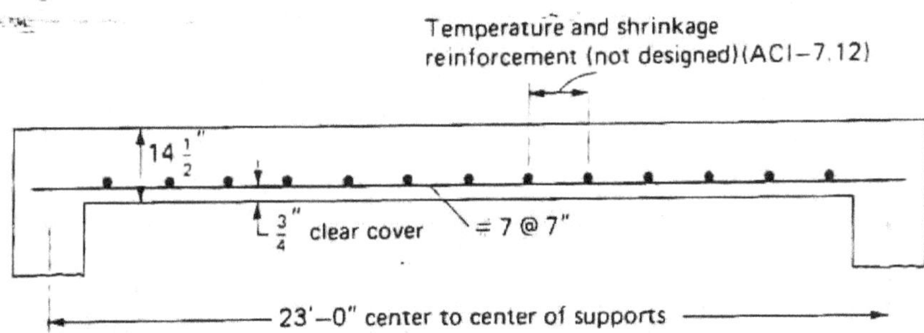

Figure 3.9.3 Slab design of Example 3.9.2.

■ **EXAMPLE 3.9.3** Compute, using SI dimensions, the nominal strength M_n of the beam of Fig. 3.9.4, 500 mm wide and 760 mm deep overall, containing 3-#35M bars in the outer layer and 2-#30M bars in the inner layer. The stirrup is #10M, and minimum clear cover is 40 mm with 25 mm between layers. Use $f'_c = 25$ MPa, $f_y = 400$ MPa, and $E_s = 200,000$ MPa.

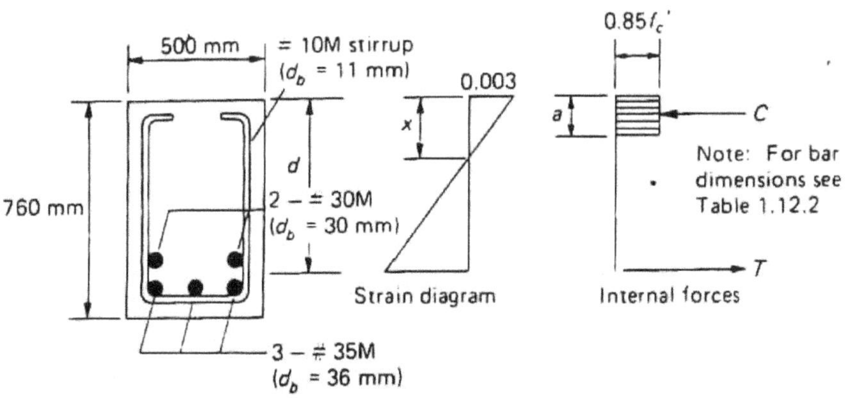

Figure 3.9.4 Computation of beam strength with metric dimensions, Example 3.9.3

Note : Usually, for main reinforcement in slabs, bars are spaced from 100 mm to 230 mm.

(f) Check strength and provide design sketch:

Use computed $d = 368$ and $A_s = 2112\ mm^2$ (per meter width of slab).

$$C = T$$

$$0.85 f_c' a b = f_y A_s \implies 0.85(20)a(1000) = 420(2112)$$

$$\implies a = 52.2\ mm.$$

$$M_n = T\left(d - \frac{a}{2}\right) = f_y A_s\left(d - \frac{a}{2}\right) = (420 \times 1000)(2112)\times 10^{-6}\left(\frac{368}{1000} - \frac{1}{2}\frac{52.2}{1000}\right)$$

$$= 303\ kN \cdot m$$

Design Strength $\phi M_n = (0.9)(303) = 273\ kN \cdot m$.

Required Strength $M_u = 263\ kN \cdot m < \phi M_n = 273\ kN \cdot m$.

$\implies$ | Use 400-mm thick slab with $\phi 22 @ 180\ mm$ spacing as shown in the figure. |

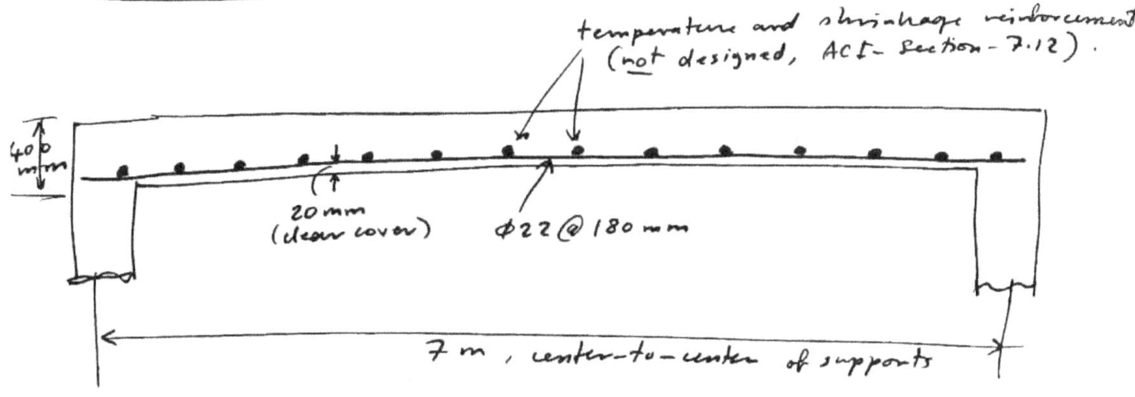

Note : Select steel transverse to the main steel for temperature and shrinkage.
(ACI Code, Section 7.12).

this is discussed later in Chapter 8.

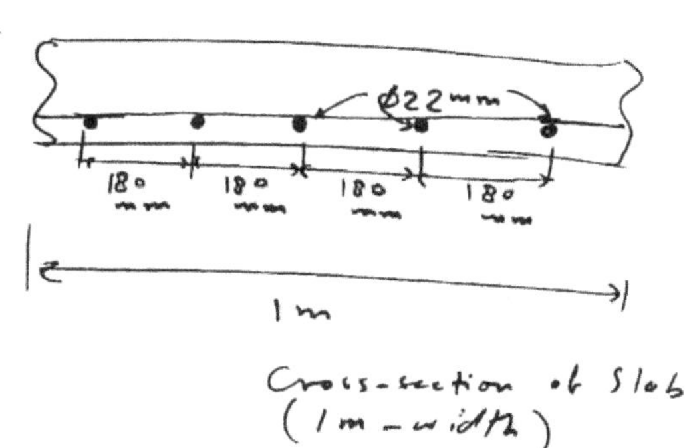

Example 7 : (Example 3.9.3 in textbook (SI Units) :

Compute, using SI dimensions, the nominal strength M_n of the beam shown in the figure, 500 mm wide and 760 mm deep overall, containing 3 - #35 M bars in the outer layer and 2 - #30 M bars in the inner layer. The stirrup is #10 M, and minimum clear cover is 40 mm with 25 mm between layers. Use $f_c' = 25$ MPa, $f_y = 400$ MPa, and $E_s = 200,000$ MPa.

Solution :

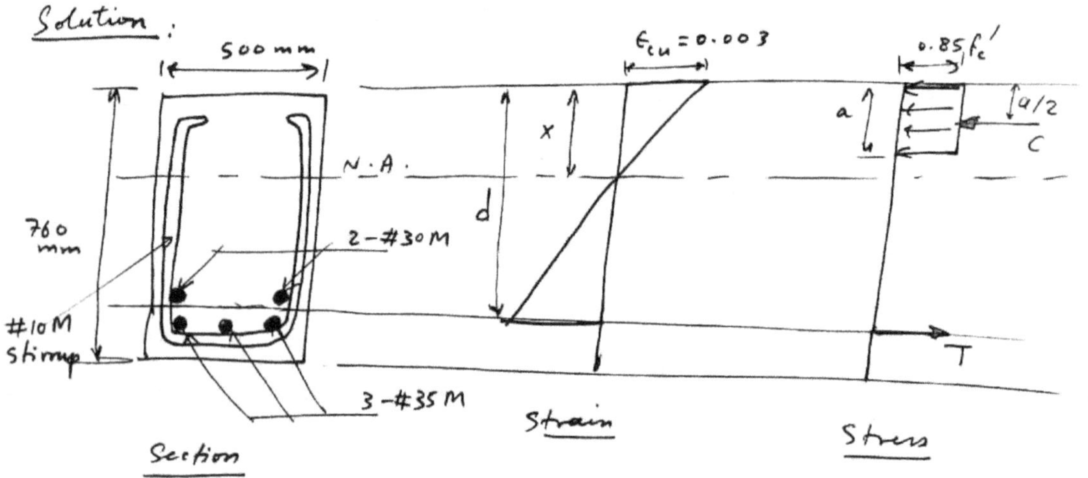

Note : These are ASTM and Canadian metric bar sizes.

$$\#10 M \longrightarrow \phi 11 \text{ mm}$$
$$\#30 M \longrightarrow \phi 30 \text{ mm}$$
$$\#35 M \longrightarrow \phi 36 \text{ mm}$$

see Table 1.12.2 page 12a in Chapter 1.

* Check if the section satisfies ACI Code requirements.

$$\epsilon_y = \frac{f_y}{E_s} = \frac{400}{200,000} = 0.002$$

Consider the balanced strain condition:

$$\frac{0.003}{x_b} = \frac{0.002}{d - x_b}$$

but
$$d = 760 - 40 - 11 - \left[\frac{(3)\left(\frac{\pi (36)^2}{4}\right)(18) + 2\left(\frac{\pi (30)^2}{4}\right)(76)}{3\left(\frac{\pi (36)^2}{4}\right) + 2\left(\frac{\pi (30)^2}{4}\right)} \right]$$

$$= 760 - 40 - 11 - 36$$

$$= 673 \text{ mm}.$$

$$\therefore \quad \frac{0.003}{x_b} = \frac{0.002}{673 - x_b} \implies 0.003(673) - 0.003\,x_b = 0.002\,x_b$$

$$\implies x_b = 404 \text{ mm}$$

For $f_c' = 25 \text{ MPa}$, we have $\beta_1 = 0.85$

$$\therefore \quad a_b = \beta_1\, x_b \implies a_b = (0.85)(404) = 343 \text{ mm}$$

$$C = T$$

$$0.85 f_c'\, a_b\, b = f_y\, A_{sb}$$

$$0.85(25)(343)(500) = 400\, A_{sb} \implies A_{sb} = 9111 \text{ mm}^2$$

$$\implies \rho_b = \frac{A_{sb}}{bd} = \frac{9111}{(500)(673)} = 0.0271 = 2.71\%$$

$$\therefore \quad \rho_{max.} = 0.75\, \rho_b = 0.75(0.0271) = 0.0203 = 2.03\%$$

$$\text{Actual } A_s = 3\left[\pi \frac{(36)^2}{4}\right] + 2\left[\pi \frac{(30)^2}{4}\right] = 4467 \text{ mm}^2$$

$$\rho = \frac{A_s}{bd} = \frac{4467}{(500)(673)} = 0.0133 = 1.33\% < 2.03\%$$

$$\text{Check also } \rho_{min} = \frac{1.4}{f_y} = \frac{1.4}{400} = 0.0035 = 0.35\%$$

$$\therefore \rho_{min} < \rho < \rho_{max}$$

The section satisfies ACI Code requirements.

* Now, we can determine M_n :

$$C = T$$

$$0.85 f_c'\, a\, b = f_y\, A_s$$

$$0.85(25)\, a\, (500) = 400(4467) \implies a = 168 \text{ mm}$$

$$M_n = T\left(d - \frac{a}{2}\right) = f_y A_s \left(d - \frac{a}{2}\right)$$

$$= (400 \times 1000)(4467 \times 10^{-6})\left(\frac{673}{1000} - \frac{1}{2}\frac{168}{1000}\right)$$

$$= 1052 \text{ kN} \cdot \text{m}$$

3.10 Investigation of Rectangular Sections in Bending with [6]

   Both Tension and Compression Reinforcement:

* **Doubly-reinforced** sections are rectangular sections with **both** tension and compression reinforcement.

Notes :
  1. Because the compressive strength of concrete is **high**, the need for compression reinforcement **to obtain adequate strength** is **not** great.

  2. When compression reinforcement is used to reduce the size of the cross-section, the following may occur:
     (a) deflections may be excessive.
     (b) there may be difficulty in placing **all** of the tension reinforcement within the width of the beam (even if two or more layers are used).
     (c) shear stresses will become high so that a large amount of shear reinforcement might be required.

  3. Compression reinforcement is mainly used to reduce creep and shrinkage deflection.

* **Analysis of Doubly-Reinforced Rectangular Sections in Bending:**

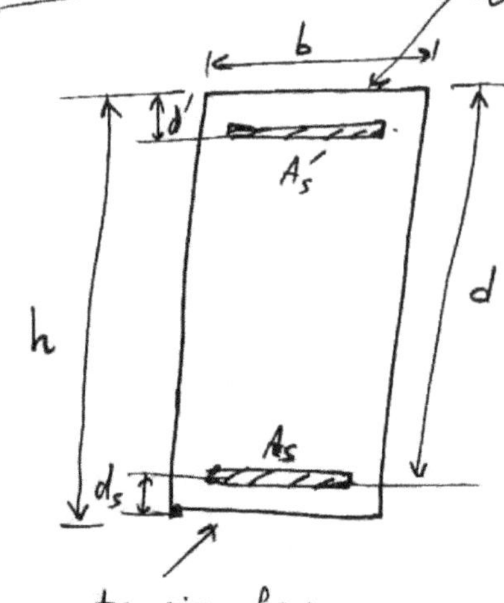

$$\rho' = \frac{A_s'}{bd} \quad : \text{ compression steel reinforcement ratio}$$

$$\rho = \frac{A_s}{bd} \quad : \text{ tension steel reinforcement ratio}$$

Analysis problems :

  Given b , d , d' , A_s , A_s' , f_c' , f_y .

  Determine M_n (nominal flexural strength)

* In case of the beam with compression steel, the ACI Code states that ρ shall not exceed $0.75\rho_b$ (just as for singly-reinforced beams).

* In addition, the ACI Code (Section 10.3.3) states that the portion of ρ_b equalized by compression reinforcement need not be reduced by the 0.75 factor.

* For doubly-reinforced rectangular section, it is easier to use and check that $\boxed{x < X_{max} = 0.75 X_b}$

* the compression steel most often yield.

Example 8 :
Determine the nominal moment strength M_n of the rectangular section shown in the figure, given $f_c' = 35$ MPa, $f_y = 415$ MPa, $b = 350$ mm, $d = 660$ mm, $d' = 75$ mm, $A_s' = 2-\phi 25$ mm bars and $A_s = 8-\phi 32$ mm bars.

Solution :

$$A_s = 8\left[\frac{\pi(32)^2}{4}\right] = 6434 \text{ mm}^2$$

$$A_s' = 2\left[\frac{\pi(25)^2}{4}\right] = 982 \text{ mm}^2$$

Yield strain in steel $\epsilon_y = \dfrac{f_y}{E_s} = \dfrac{415}{200,000} = 0.002075$

(a) Determine A_{sy} in the <u>compression steel yield condition.</u>

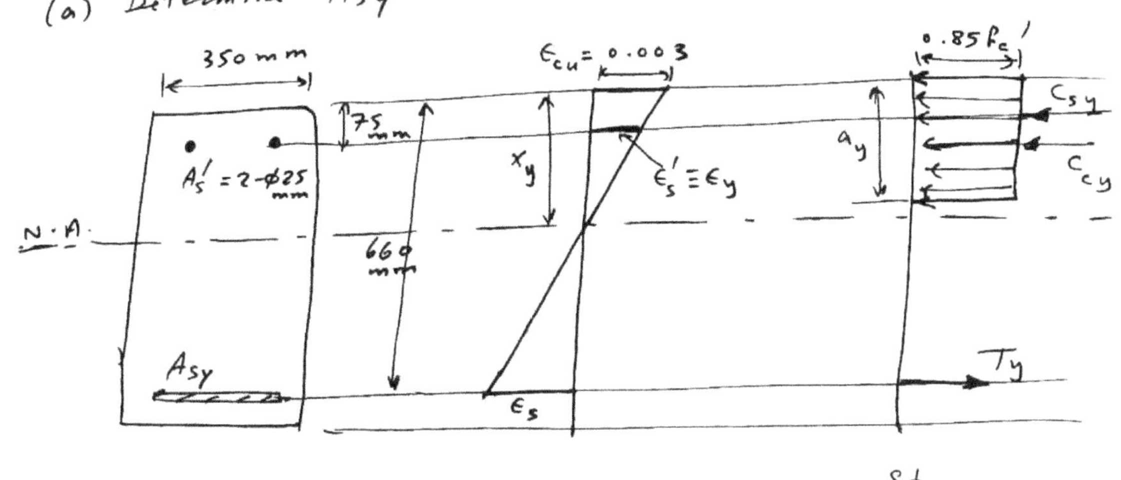

By the similar triangles (top two triangles). with $\epsilon_s' = \epsilon_y$.

$$\frac{0.002075}{x_y - 75} = \frac{0.003}{x_y}$$

$$0.002075\, x_y = 0.003 x_y - 0.003(75)$$

$$\Rightarrow x_y = 243 \text{ mm}.$$

For $f_c' = 35$ MPa, $\beta_1 = 0.85 - 0.08\left(\frac{35-30}{10}\right) = 0.81$.

$$\therefore a_y = \beta_1 x_y = (0.81)(243) = 197 \text{ mm}.$$

$$C_{cy} = 0.85 f_c' a_y b = 0.85(35\times1000)\left(\frac{197}{1000}\right)\left(\frac{350}{1000}\right) = 2051 \text{ kN}.$$

$$C_{sy} = (f_y - 0.85 f_c')A_s' = \left(415 - 0.85(35)\right)\times1000\times982\times10^{-6} = 378 \text{ kN}.$$

By equilibrium: $\xrightarrow{+} \Sigma F_x = 0:$ $T_y - C_{cy} - C_{sy} = 0$

$$\Rightarrow T_y = C_{cy} + C_{sy} = 2051 + 378 = 2429 \text{ kN}.$$

$$T_y = f_y A_{sy} \Rightarrow 2429 = (415\times1000) A_{sy}$$

$$\Rightarrow A_{sy} = 0.005853 \text{ m}^2$$
$$= 5853 \text{ mm}^2.$$

Actual $A_s = 6434 \text{ mm}^2 > A_{sy} = 5853 \text{ mm}^2$.

$$\Rightarrow \text{Compression steel yields.}$$

(b) Determine $A_{s_{max}}$: Use the balanced strain condition.

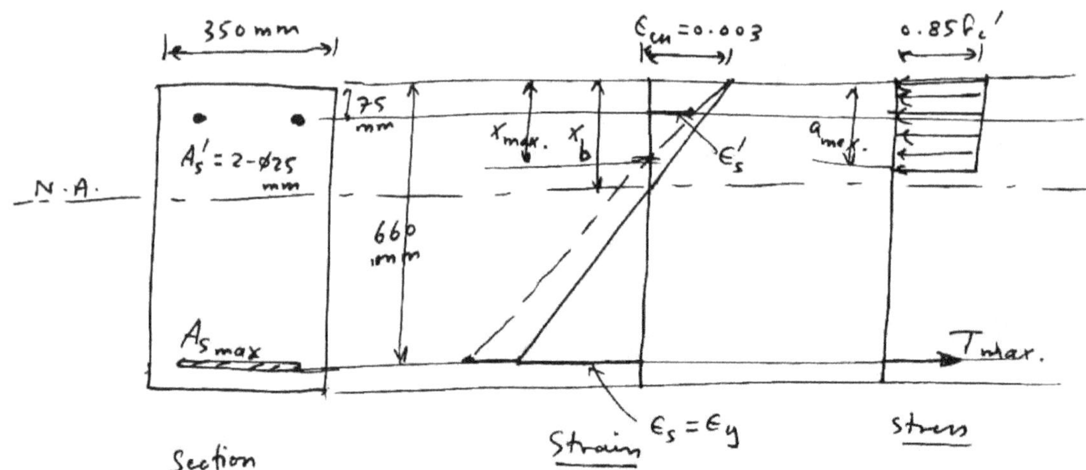

From similar triangles, with $\epsilon_s = \epsilon_y$

$$\frac{0.003}{x_b} = \frac{0.002075}{660 - x_b}$$

$$0.003(660) - 0.003 x_b = 0.002075 x_b$$

$$\Rightarrow \quad x_b = 390 \text{ mm}.$$

$$x_{max} = 0.75 x_b = 0.75(390) = 292.5 \text{ mm}.$$

$$a_{max} = \beta_1 x_{max} = (0.81)(292.5) = 237 \text{ mm}.$$

$$C_{c_{max}} = 0.85 f_c' a_{max} b = (0.85)(35 \times 1000)\left(\frac{237}{1000}\right)\left(\frac{350}{1000}\right) = 2468 \text{ kN}$$

From similar triangles (with x_{max}) (top two triangles):

$$\frac{\epsilon_s'}{292.5 - 75} = \frac{0.003}{292.5} \quad \Rightarrow \quad \epsilon_s' = 0.00223 > \epsilon_y = 0.00207.$$

$\therefore$ Compression steel yields.

$$C_s = (f_y - 0.85 f_c') A_s' = [415 - 0.85(35)] \times 1000 \times 982 \times 10^{-6}$$
$$= 378 \text{ kN}.$$

$$T_{max} = C_{c_{max}} + C_s = 2468 + 378 = 2846 \text{ kN}.$$

$$\Rightarrow \quad T_{max} = f_y A_{s\,max}$$
$$2846 = (415 \times 1000) A_{s\,max} \quad \Rightarrow \quad A_{s\,max} = 0.006858 \text{ m}^2$$
$$= 6858 \text{ mm}^2.$$

$$\therefore \text{ Actual } A_s = 6434 \text{ mm}^2 < A_{s\,max} = 6858 \text{ mm}^2.$$

<u>Note</u>: Calculate A_{sb} for this case and compare:

$$a_b = \beta_1 x_b = (0.81)(390) = 316 \text{ mm}.$$
$$C_{c_b} = 0.85 f_c' a_b b = 0.85(35 \times 1000)\left(\frac{316}{1000}\right)\left(\frac{350}{1000}\right)$$
$$= 3290 \text{ kN}.$$

$$\therefore \quad T_b = C_{c_b} + C_s = 3290 + 378 = 3668 \text{ kN}.$$

$$T_b = f_y A_{sb} \quad \Rightarrow \quad 3668 = (415 \times 1000) A_{sb} \Rightarrow A_{sb} = 0.008839 \text{ m}^2$$
$$= 8839 \text{ mm}^2.$$

$$\therefore \quad \frac{A_{s\,max}}{A_{sb}} = \frac{6858}{8839} \times 100\% = 78\% > 75\%.$$
$$\text{(o.k. because of compression steel).}$$

(c) Determine M_n from the actual condition:

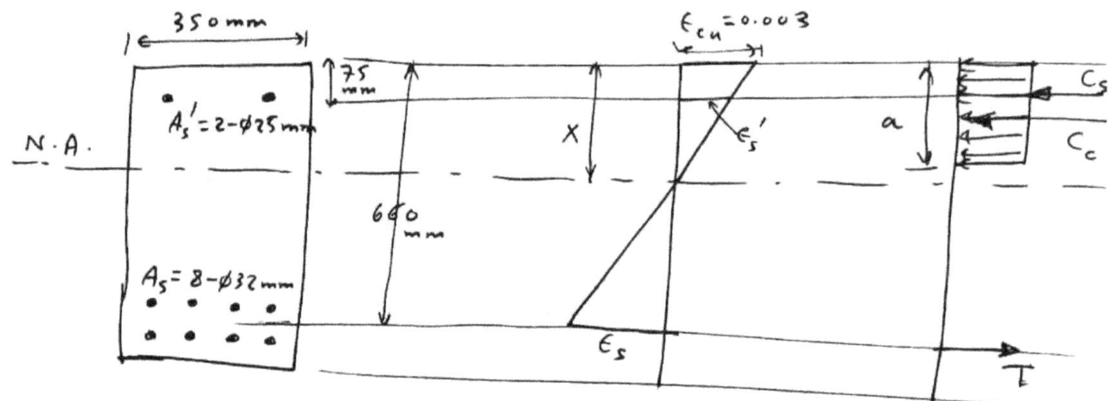

$$C_c + C_s = T \qquad \text{(equilibrium)}$$

$$0.85 f_c' a b + (f_y - 0.85 f_c') A_s' = f_y A_s$$

$$0.85 (35) a (350) + [415 - 0.85 (35)](982) = 415 (6434)$$

$$\Rightarrow a = 220 \text{ mm}.$$

$$a = \beta_1 x \Rightarrow 220 = 0.81 x \Rightarrow x = 272 \text{ mm}.$$

Check that $x < 0.75 \, x_b = 0.75 (390) = 292.5 \text{ mm}$

$$\text{o.k.}$$

Check ϵ_s' :
From similar triangles: $\dfrac{\epsilon_s'}{x - d'} = \dfrac{\epsilon_{cu}}{x}$

$$\frac{\epsilon_s'}{272 - 75} = \frac{0.003}{272} \Rightarrow \epsilon_s' = 0.00217 > \epsilon_y = 0.00207.$$

$\therefore$ Compression steel yields.

$$M_n = C_c \left(d - \frac{a}{2}\right) + C_s (d - d') \qquad \text{by taking moments about } T$$

$$= 0.85 f_c' a b \left(d - \frac{a}{2}\right) + (f_y - 0.85 f_c')(d - d') A_s'$$

$$= 0.85 (35 \times 1000) \left(\frac{220}{1000}\right)\left(\frac{350}{1000}\right)\left(\frac{660}{1000} - \frac{1}{2}\frac{220}{1000}\right)$$

$$\quad + [415 - 0.85 (35)] \times 1000 \left(\frac{660}{1000} - \frac{75}{1000}\right)(982 \times 10^{-6})$$

$$= 1481 \text{ kN} \cdot \text{m}$$

<u>Example 9</u>:

Repeat the solution for Example 8, except that A_s is 4-ϕ36 mm bars instead of 8-ϕ32 mm bars. Also, determine the percentage increase in the strength of the beam over the same beam without compression steel.

<u>Solution</u>: $A_s = 4\left(\dfrac{\pi (36)^2}{4}\right) = 4072 \text{ mm}^2$.

(a) Determine A_{sy} in the compression steel yield condition.

From Example-8-(a), $A_{sy} = 5853 \text{ mm}^2$

Actual $A_s = 4072 \text{ mm}^2 < A_{sy} = 5853 \text{ mm}^2$.

∴ Compression steel does not yield.

(b) Determine M_n from the actual condition:

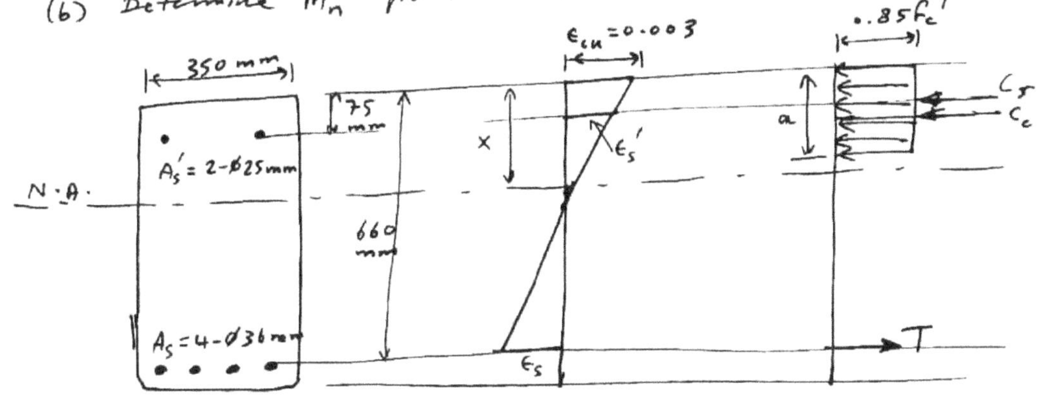

Since compression steel does not yield, we use f_s', <u>not</u> f_y.

$$\frac{\epsilon_s'}{0.003} = \frac{x-75}{x}$$

$$\Rightarrow \quad \epsilon_s' = 0.003\left(\frac{x-75}{x}\right)$$

$$\therefore \quad f_s' = E_s\,\epsilon_s' = (200,000)(0.003)\left(\frac{x-75}{x}\right)$$

$$= 600\left(\frac{x-75}{x}\right), \text{ in } \frac{MPa.}{(x \text{ in mm})}.$$

<u>Equilibrium</u> : $\xrightarrow{+} \Sigma F_x = 0$:

$$C_c + C_s = T$$

$$0.85 f_c'\,ab + (f_s' - 0.85 f_c')A_s' = f_y A_s$$

$$\text{but} \quad a = \beta_1 x = 0.81 x .$$

$$0.85(35)(0.81x)(350) + \left[600\left(\frac{x-75}{x}\right) - 0.85(35) \right](982) = 415(4072)$$

$$\Rightarrow \quad 8434.125 \, x^2 + 589,200 \, (x-75) - 2921.5 \, x = 1,689,800 \, x$$

$$8434.125 \, x^2 - 1129,814.5 \, x - 75(589,200) = 0$$

$$\therefore \quad x = \frac{1,129,814.5 \pm \sqrt{(1,129,814.5)^2 - 4(8434.125)(-75)(589,200)}}{2(8434.125)}$$

$$= 166 \text{ mm} .$$

$$\therefore \quad a = \beta_1 x = 0.81(166) = 134 \text{ mm} .$$

$$C_c = 0.85 f_c' a b = 0.85(35 \times 1000)\left(\frac{134}{1000}\right)\left(\frac{350}{1000}\right) = 1395 \text{ kN} .$$

$$\epsilon_s' = 0.003\left(\frac{166-75}{166}\right) = 0.00164 < \epsilon_y = 0.002075 .$$

$$\therefore \quad \text{Compression steel does } \underline{\text{not}} \text{ yield, as assumed.}$$

$$C_s = (f_s' - 0.85 f_c') A_s'$$

$$= (E_s \epsilon_s' - 0.85 f_c') A_s'$$

$$= \left[(200,000)(0.00164) - (0.85)(35) \right] \times 1000 \times 982 \times 10^{-6}$$

$$= 293 \text{ kN} .$$

$$\Rightarrow \quad C_c + C_s = 1395 + 293 = 1688 \text{ kN} .$$

$$\underline{\text{Check}}: \quad T = f_y A_s = (415 \times 1000) \times 4072 \times 10^{-6}$$

$$= 1690 \text{ kN}$$

$$\approx 1688 \text{ kN} \quad \text{o.k.}$$

$$\therefore \quad M_n = C_c\left(d - \frac{a}{2}\right) + C_s(d - d')$$

$$= (1395)\left(\frac{660}{1000} - \frac{1}{2}\frac{134}{1000}\right) + (293)\left(\frac{660}{1000} - \frac{75}{1000}\right)$$

$$= \underline{\underline{999}} \text{ kN} \cdot \text{m} .$$

(c) For a singly-reinforced beam
    (without compression steel):

$$\rho_b = \frac{0.85 f_c'}{f_y} \beta_1 \left(\frac{600}{600+f_y}\right) = \frac{0.85(35)}{415}(0.81)\left(\frac{600}{600+415}\right)$$

$$= 0.0343$$

$$\therefore \rho_{max.} = 0.75\,\rho_b = 0.75\,(0.0343) = 0.0257 = 2.57\%.$$

$$\text{Actual } \rho = \frac{A_s}{bd} = \frac{4072}{350\times660} = 0.0176 \quad \begin{array}{l} = 1.76\% \\ < 2.57\% \end{array}$$

Now, compute M_n from the actual condition:

$$C = T$$
$$0.85 f_c'\, a\, b = f_y A_s$$
$$0.85\,(35)\, a\,(350) = (415)(4072)$$

$$\Rightarrow \quad a = 162\ mm.$$

$$\therefore \quad M_n = T\left(d - \frac{a}{2}\right) = f_y A_s\left(d - \frac{a}{2}\right)$$
$$= (415\times1000)(4072\times10^{-6})\left(\frac{660}{1000} - \frac{1}{2}\frac{162}{1000}\right)$$

$$= \underline{\underline{978}}\ kN\cdot m$$

Increase in nominal strength $= \dfrac{999-978}{978}\times100\% = 2.1\%.$

<u>Note</u>: It is clear that compression steel was <u>not</u> used
because the strength was inadequate. It was used for
deflection control.

3.11 Criterion for the Compression Steel Yield Condition:

* In a doubly-reinforced section, the criterion to ensure
that the compression steel yields when the nominal
strength is reached is:

$$\rho - \rho'\left(1 - \frac{0.85 f_c'}{f_y}\right) \geq 0.85\,\beta_1\left(\frac{f_c' d'}{f_y d}\right)\left(\frac{87,000}{87,000 - f_y}\right).$$

$$f_c', f_y \text{ in psi.}$$

In SI units,

$$\rho - \rho'\left(1 - \frac{0.85 f_c'}{f_y}\right) \geq 0.85\,\beta_1\left(\frac{f_c'\,d'}{f_y\,d}\right)\left(\frac{600}{600 - f_y}\right),$$

$$f_c',\ f_y \ \text{in MPa}.$$

where $\rho' = \dfrac{A_s'}{bd}$ (compression reinforcement ratio).

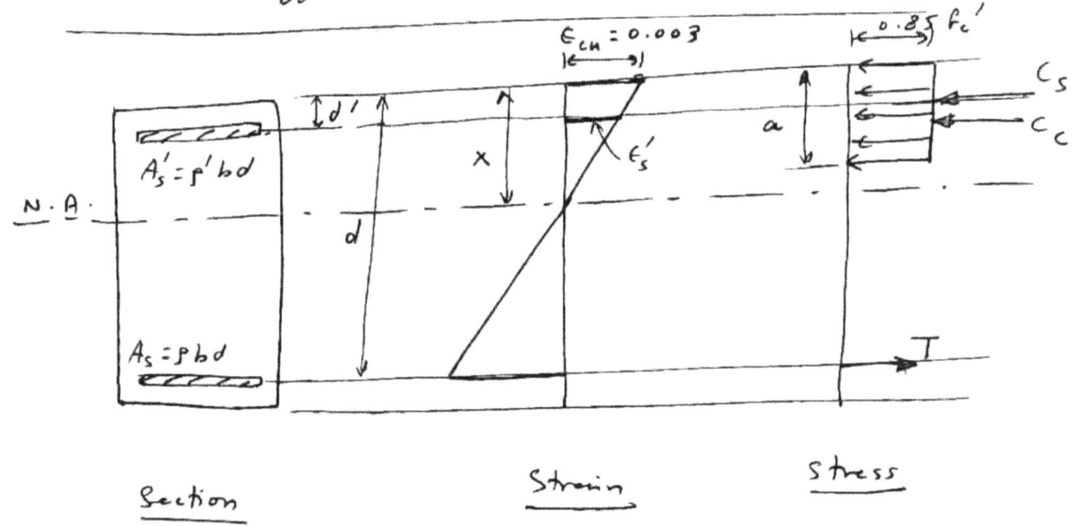

Criterion for yielding of the compression steel is:

$$\epsilon_s' \geq \epsilon_y$$

$$T = f_y A_s = \rho b d\, f_y\ .$$

$$C_c = 0.85 f_c'\,ab = 0.85 f_c'\,\beta_1 x\, b$$

$$C_s = (f_y - 0.85 f_c')A_s' = (f_y - 0.85 f_c')\rho' bd$$

Equilibrium $\Rightarrow$ $\qquad C_c + C_s = T$

$$0.85 f_c'\,\beta_1 x\, b + (f_y - 0.85 f_c')\rho' bd = \rho b d\, f_y$$

$$0.85 f_c'\,\beta_1 x\, b = \rho b d\, f_y - f_y\left(1 - \frac{0.85 f_c'}{f_y}\right)\rho' bd$$

$$0.85 f_c'\,\beta_1 x\, b = bd\, f_y\left[\rho - \rho'\left(1 - \frac{0.85 f_c'}{f_y}\right)\right]$$

$$x = \frac{f_y d}{0.85 f_c'\,\beta_1}\left[\rho - \rho'\left(1 - \frac{0.85 f_c'}{f_y}\right)\right]$$

From similar triangles:

$$\frac{\epsilon_s'}{0.003} = \frac{x-d'}{x} \quad\Longrightarrow\quad \epsilon_s' = 0.003\left(\frac{x-d'}{x}\right)$$

For yielding of compression steel, we have the condition:

$$\epsilon_s' \geq \epsilon_y$$

$$0.003\left(\frac{x-d'}{x}\right) \geq \frac{f_y}{200,000} \quad,\qquad f_y \text{ in MPa}.$$

$$\Longrightarrow \quad \frac{x-d'}{x} \geq \frac{f_y}{600}$$

$$1 - \frac{d'}{x} \geq \frac{f_y}{600} \quad\Longrightarrow\quad \frac{d'}{x} \leq 1 - \frac{f_y}{600} = \frac{600-f_y}{600}$$

$$\Longrightarrow \quad x \geq \left(\frac{600}{600-f_y}\right)d'$$

but $\quad x = \dfrac{f_y d}{0.85 f_c' \beta_1}\left[\rho - \rho'\left(1 - \dfrac{0.85 f_c'}{f_y}\right)\right]$

$$\therefore \quad \frac{f_y d}{0.85 f_c' \beta_1}\left[\rho - \rho'\left(1 - \frac{0.85 f_c'}{f_y}\right)\right] \geq \left(\frac{600}{600-f_y}\right)d'$$

$$\Longrightarrow \quad \boxed{\rho - \rho'\left(1 - \frac{0.85 f_c'}{f_y}\right) \geq \frac{0.85 f_c' \beta_1 d'}{f_y d}\left(\frac{600}{600-f_y}\right)}\quad,$$

$$f_c', f_y \text{ in MPa}.$$

3.12 Design of Rectangular Sections in Bending with Both Tension and Compression Reinforcement:

* the primary reason for using compression reinforcement is to reduce long-time deflection due to creep and shrinkage.

* Use the two equilibrium equations:

$$\boxed{\begin{array}{l} C_c + C_s = T \qquad\qquad (1) \\[2ex] M_n = C_c\left(d - \frac{a}{2}\right) + C_s\left(d - d'\right) \qquad (2) \end{array}}$$

where $a = \beta_1 x$.

* In the design of doubly-reinforced rectangular sections,
  check the following :

    (1) $x < 0.75\, X_b$.

    (2)   If the compression steel is assumed to yield,
  check that it actually yields (use f_y) .

    (3)   If the compression steel does <u>not</u> yield,
  reformulate the problem using f_s', not f_y .

* If deflection control is important, take $x < 0.375\, X_b$.

<u>Example 10</u> :

    Determine the A_s and A_s' required to carry a service
live load moment of 520 kN·m and a service dead load
moment of 267 kN·m , using $b = 350$ mm , $d = 660$ mm,
$d' = 75$ mm , $f_c' = 35$ MPa , $f_y = 415$ MPa , and the
ACI Code , as shown in the figure .

<u>Solution</u> :

(a) Determine required strength M_u :

$$M_u = 1.4 M_d + 1.7 M_\ell$$
$$= 1.4\,(267) + 1.7\,(520)$$
$$= 1257.8 \text{ kN·m}$$

$$\text{Required } M_n = \frac{M_u}{\phi} = \frac{1257.8}{0.9} = 1398 \text{ kN·m}$$

(b) Determine the maximum strength and reinforcement
allowed by the ACI Code for a <u>singly-reinforced</u> section.

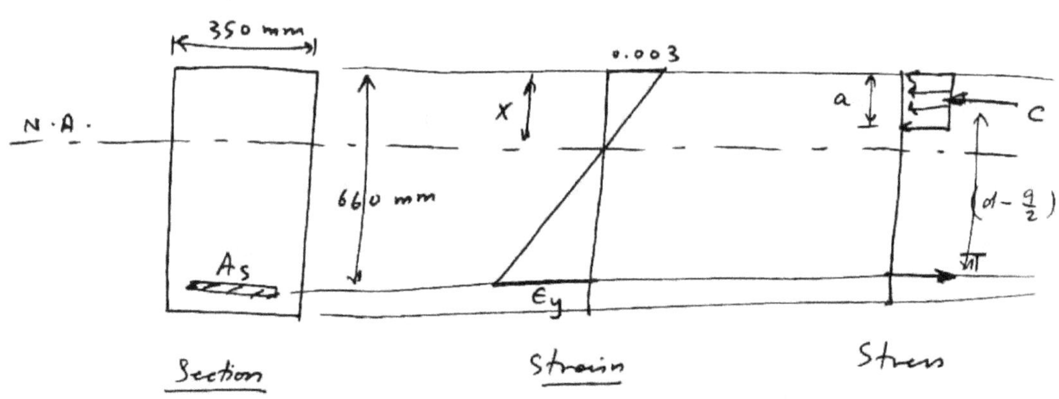

Balanced Strain Condition: $\varepsilon_y = \dfrac{f_y}{E_s} = \dfrac{415}{200,000} = 0.002075$.

$$\frac{0.003}{X_b} = \frac{0.002075}{660 - X_b}$$

$$0.003(660) - 0.003 X_b = 0.002075 X_b$$

$$\Longrightarrow \quad X_b = 390 \text{ mm}.$$

$\therefore \quad X_{max} = 0.75 X_b = 0.75(390) = 292.5 \text{ mm}$

$$C_{max} = 0.85 f_c' a_{max} b = 0.85 f_c'(0.81 X_{max}) b,$$

$$\text{Here} \quad \beta_1 = 0.81 \quad \text{for} \quad f_c' = 35 \text{ MPa}.$$

$$\therefore \quad C_{max} = 0.85 (35 \times 1000)(0.81)\left(\frac{292.5}{1000}\right)\left(\frac{350}{1000}\right) = 2467 \text{ kN}.$$

$$C_{max} = T_{max} = f_y A_s$$

$$2467 = (415 \times 1000) A_s \Longrightarrow A_s = 0.005945 \text{ m}^2 = 5945 \text{ mm}^2.$$

$$\therefore \quad M_{n\,(max)} = C_{max}\left(d - \frac{a}{2}\right) = 2467\left(\frac{660}{1000} - \frac{1}{2}\frac{(0.81)(292.5)}{1000}\right)$$

$$= 1336 \text{ kN} \cdot \text{m}$$

<u>Note</u> :

Required $M_n = 1398 \text{ kN} \cdot \text{m} > M_{n\,max} = 1336 \text{ kN} \cdot \text{m}$

$\Longrightarrow$ Compression reinforcement is needed for strength.

(c) Determine the minimum compression reinforcement required:

Keep $X_{max} = 0.75 X_b = 292.5 \text{ mm}$.

Take $M_{n_c} = 1336 \text{ kN} \cdot \text{m}$.

$\therefore \quad M_{n_s} = M_{n\,required} - M_{n_c} = 1398 - 1336 = 62 \text{ kN} \cdot \text{m}$

$$M_{n_s} = C_s(d - d') \qquad (\text{Required additional strength}).$$

$$62 = C_s\left(\frac{660}{1000} - \frac{75}{1000}\right) \Longrightarrow C_s = 106 \text{ kN}.$$

Check if compression steel yields:

$$\frac{\epsilon_s'}{x-75} = \frac{0.003}{x}$$

$$\frac{\epsilon_s'}{292.5-75} = \frac{0.003}{292.5}$$

$$\Rightarrow \quad \epsilon_s' = 0.00223 > \epsilon_y = \frac{f_y}{200,000} = \frac{415}{200,000} = 0.002075$$

$\therefore$ Compression steel yields $\Rightarrow$ use f_y in the formula.

$$C_s = (f_y - 0.85 f_c') A_s'$$

$$106 = [415 - 0.85(35)] \times 1000 \, A_s'$$

$$\Rightarrow A_s' = 0.0002751 \, m^2$$

$$= \underline{\underline{275 \, mm^2}}.$$

$$\Rightarrow \quad T = C_{c_{max}} + C_s = 2467 + 106 = 2573 \, kN.$$

$$T = f_y A_s$$

$$2573 = (415 \times 1000) A_s \Rightarrow A_s = 0.0062 \, m^2$$

$$= \underline{\underline{6200 \, mm^2}}.$$

<u>Note</u>: A_s given above is exactly the <u>maximum</u> tension reinforcement permitted by the ACI Code.

<u>Example 11</u>:

Redesign the section of Example 10 so that the actual neutral axis location is at 0.375 of that in the balanced strain condition. Assume that the purpose is for deflection control. From example 10, use $M_u = 1257.8 \, kN \cdot m$

<u>Solution</u>:

(a) $\qquad x = 0.375 x_b = 0.375(390) = 146 \, mm$.

$$a = 0.81 x = 0.81(146) = 118 \, mm.$$

$$C_c = 0.85 f_c' a b = 0.85(35 \times 1000)\left(\frac{118}{1000}\right)\left(\frac{350}{1000}\right) = 1229 \, kN.$$

$$T_1 = C_c = 1229 \, kN.$$

$$M_{n_c} = C_c\left(d - \frac{a}{2}\right) = 1229\left(\frac{660}{1000} - \frac{118}{2(1000)}\right) = 738 \text{ kN}\cdot\text{m}$$

Let A_{sc} : part of tension steel to match the concrete in compression

$$T_1 = f_y A_{sc} \implies 1229 = 415 \times 1000 \, A_{sc}$$

$$\implies A_{sc} = 0.002961 \text{ m}^2$$

$$= 2961 \text{ mm}^2 .$$

(b) Determine steel requirements for both faces of the beam:

$$M_{n_{required}} = \frac{M_u}{\phi} = \frac{1257.8}{0.9} = 1398 \text{ kN}\cdot\text{m}$$

$$M_{n_{required}} = M_{n_c} + M_{n_s}$$

$$1398 = 738 + M_{n_s} \implies M_{n_s} = 660 \text{ kN}\cdot\text{m}$$

However, $\quad M_{n_s} = C_s(d - d')$

$$660 = C_s\left(\frac{660}{1000} - \frac{75}{1000}\right)$$

$$\implies C_s = 1128 \text{ kN} .$$

Check if compression steel yields:

$$\frac{0.003}{146} = \frac{\varepsilon_s'}{146 - 75}$$

$$\implies \varepsilon_s' = 0.00146 < \varepsilon_y = 0.002075$$

$\therefore$ Compression steel does <u>not</u> yield.

$\therefore$ Use f_s', <u>not</u> f_y for compression steel.

$$f_s' = E_s \varepsilon_s' = (200,000)(0.00146) = 292 \text{ MPa} .$$

$$C_s = (f_s' - 0.85 f_c') A_s'$$

$$1128 = \left[292 - 0.85(35)\right] \times 1000 \, A_s'$$

$$\implies A_s' = 0.004301 \text{ m}^2$$

$$= 4301 \text{ mm}^2 .$$

$$T_2 = C_s = f_y A_{ss} \implies 1128 = (415 \times 1000) A_{ss}$$

$$\Rightarrow A_{ss} = 0.002718 \ m^2$$
$$= 2718 \ mm^2.$$

$$\therefore \ A_s = A_{sc} + A_{ss} = 2961 + 2718 = 5679 \ mm^2.$$

Required steel areas are:
$$A_s = 5679 \ mm^2.$$
$$A_s' = 4301 \ mm^2.$$

Select $10-\phi28\ mm$ (tension reinforcement in two layer and $4-\phi36\ mm$ (compression reinforcement).

Check if selected reinforcement fits within the beam width

Bar clearance in tension reinforcement

$$= \frac{350 - 2(40) - 2(10) - 5(28)}{4} = 27.5 \ mm \approx 28 \ mm$$
$$o.k.$$

Bar clearance in compression reinforcement

$$= \frac{350 - 2(40) - 2(10) - 4(36)}{3} = 35.3 \ mm > 36 \ m$$
$$o.k.$$
$$[0.7 \ mm \ difference \ is \ o.k$$

$$\therefore \ Selected \ areas \ are \ \ A_s = 6158 \ mm^2.$$
$$A_s' = 4072 \ mm^2.$$

<u>Note</u>: A_s' selected is less than A_s' required because $4-\phi36\ mm$ is the maximum steel that will fit in one layer.

(c) Check the strength of the section:

Since A_s' selected is less than that computed, we ne to find the new location of the neutral axis x:

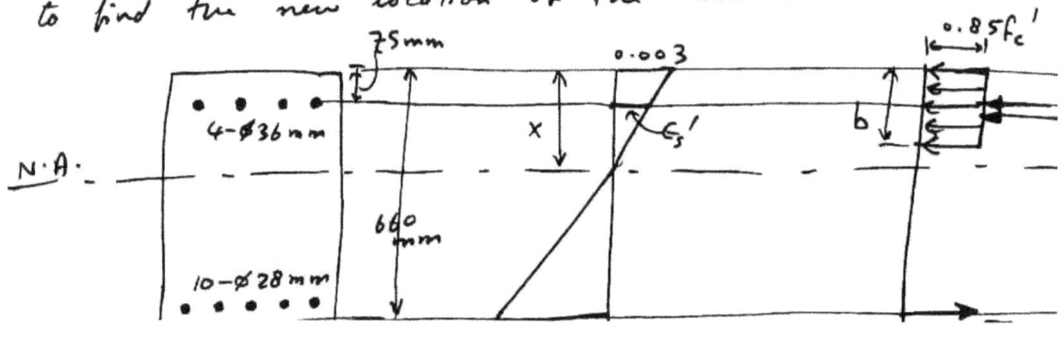

Since f_s' depends on x, we will get a quadratic equation in x:

$$\frac{\epsilon_s'}{0.003} = \frac{x-75}{x} \implies \epsilon_s' = 0.003\left(\frac{x-75}{x}\right).$$

$$\therefore \ f_s' = E_s\,\epsilon_s' = (200{,}000)(0.003)\left(\frac{x-75}{x}\right)$$

$$= 600\left(\frac{x-75}{x}\right), \qquad f_s' \text{ in MPa}.$$

Equilibrium: $\qquad C_c + C_s = T$

$$0.85 f_c'\,ab + (f_s' - 0.85 f_c')\,A_s' = f_y A_s$$

but $a = \beta_1 x = 0.81 x$

$$\implies 0.85 f_c'(0.81x)\,b + (f_s' - 0.85 f_c')\,A_s' = f_y A_s$$

$$0.85(35)(0.81x)(350) + \left[600\left(\frac{x-75}{x}\right) - 0.85(35)\right](4072) =$$

$$\qquad\qquad\qquad + \ 415(6158)$$

$$\implies 8434.125\,x^2 + 2{,}443{,}200(x-75) - 121{,}142x$$
$$= 2{,}555{,}570\,x$$

$$\implies 8434.125\,x^2 - 233512 \ x - 75(2{,}443{,}200) = 0$$

$$\implies x = \frac{233512 \pm \sqrt{(-233512)^2 - 4(8434.125)(2{,}443{,}200)(-75)}}{2(8434.125)}$$

$$\implies x = 162 \text{ mm}.$$

$$\therefore \quad a = \beta_1 x = (0.81)(162) = 131 \text{ mm}.$$

$$C_c = 0.85 f_c'\,a\,b = 0.85(35\times1000)\left(\frac{131}{1000}\right)\left(\frac{350}{1000}\right) = 1364 \text{ kN}.$$

$$C_s = (f_s' - 0.85 f_c')\,A_s'$$

$$= \left[600\left(\frac{162-75}{162}\right) - 0.85(35)\right]\times1000\,(4072\times10^{-6})$$

$$= 1191 \text{ kN}.$$

Check that the compression steel does not yield:

$$\varepsilon_s' = 0.003 \left(\frac{162 - 75}{162} \right) = 0.00161 < \varepsilon_y = 0.002075 \quad \text{o.k.}$$

$$\Rightarrow M_n = C_c \left(d - \frac{a}{2} \right) + C_s (d - d')$$

$$= 1364 \left(\frac{660}{1000} - \frac{1}{2} \frac{131}{1000} \right) + 1191 \left(\frac{660}{1000} - \frac{75}{1000} \right)$$

$$= 1508 \text{ kN} \cdot \text{m}$$

Design Strength $\phi M_n = (0.9)(1508) = 1357$ kN·m.

$$M_u = 1257.8 \text{ kN} \cdot \text{m} < \phi M_n = 1357 \text{ kN} \cdot \text{m}$$

3.13 Non-Rectangular Beams:

* The shape of the compression zone dictates the formulas to be used.

* The Whitney rectangular stress block can be used for non-rectangular beams.

* $A_{s_{max}}$ is obtained by setting $x = x_{max} = 0.75 X$
  (or set $A_s = A_{s\,max} = 0.75 A_{s_b}$).

* When a section has the greater compression area near the extreme fiber, (as for a T-section), the tension steel strain will be greater than for the rectangular beam i.e. $\left(x_{max} < 0.75 x_b \right)$.

* When a section has the greater compression area near the neutral axis, the tension strain in steel achieved will be less than for the rectangular beam; i.e. $\left(x_{max} > 0.75 x_b \right)$.

* Examples on Non-Rectangular Beams are provided in the Design Sessions.

Chapter 4

Rectangular Sections in Bend
Under Service Load C

4.1 General Introduction:

(1) <u>Strength Design Method</u> (studied in Chapt

characterized by.
1. load factors.
2. reduction factors.
3. nonlinear stress-strain relat
(rectangular Whitney str

(2) <u>Working Stress Method</u> (to be studied in C

characterized by:
1. service loads.
2. allowable working stresses
3. linear stress-strain relati

---

* The <u>working stress method</u> is referred to
as the <u>alternate design method.</u>

* This method is involved with <u>serviceabilit</u>
performance under service loads (ie. work

* Satisfactory performance is defined in term.
1. <u>deflection</u> within acceptable limit
nonstructural elements such as wall.
ceilings are not damaged.
2. <u>cracking</u> controlled to prevent la

(3) Concrete does not take tension (concrete crac
under tension).

(4) No slip occurs between the steel bars and
surrounding concrete during the development
forces in the bars.

<u>Note</u> : * Only assumption (2) above is unique to
stress method.
* This is reasonably accurate for stresses bel

4.3 Modulus of Elasticity Ratio n : (Modular !

For steel, $E_s = 200,000$ MPa .

For concrete, E_c varies with $\sqrt{f_c'}$ and w_c

* Modular ratio $\boxed{n = \dfrac{E_s}{E_c}}$ is taken as the n

number, but not less than 6 . [AC1 Code

f_c' (psi)	n
3000	9
3500	8.5
4000	8
4500	7.5
5000	7
6000	6.5

f_c' (MPa)	n
20	9
25	8
30	7.5
35	7
40	6.5

4.4. Equilibrium Conditions: $\Rightarrow \Sigma F_v = 0$. $\circlearrowleft \Sigma M$

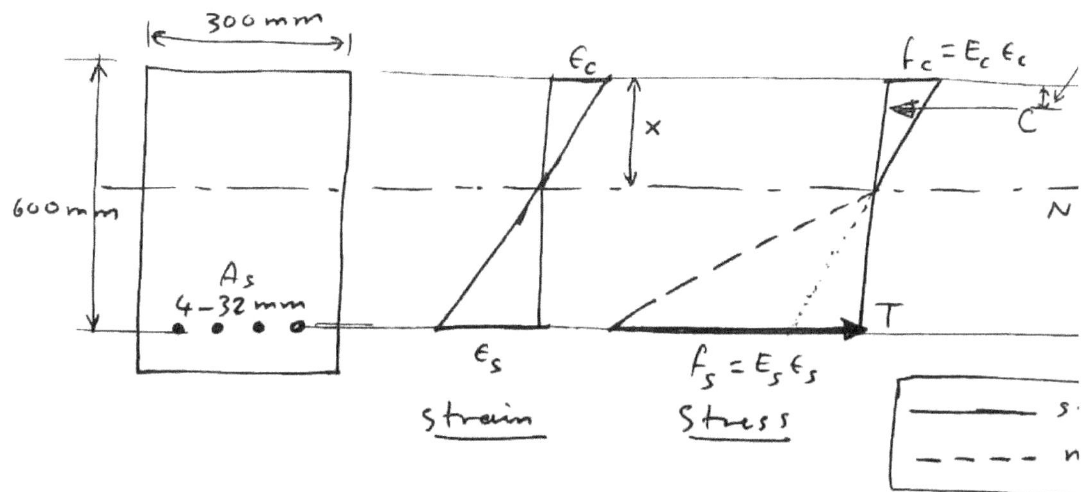

Compressive Force $\quad C = \frac{1}{2} f_c x b = \frac{1}{2} f_c \times (300) = 150$

Tensile Force $\quad T = f_s A_s = f_s \left(4 \times \frac{\pi (32)^2}{4} \right) = 3215$

$\xrightarrow{+} \Sigma F_x = 0: \quad C = T$

$$150 f_c x = 3215 f_s \implies \frac{f_s}{f_c} = \frac{150 x}{3215} \quad —$$

From the strain diagram (by similar triangles)

$$\frac{\epsilon_s}{\epsilon_c} = \frac{d - x}{x} = \frac{600 - x}{x}$$

but $\quad \epsilon_s = f_s / E_s \quad$ and $\quad \epsilon_c = f_c / E_c$

$$\implies \frac{f_s}{E_s} \cdot \frac{E_c}{f_c} = \frac{600 - x}{x}$$

$$\frac{f_s}{f_c} \left(\frac{E_c}{E_s} \right) = \frac{600 - x}{x}$$

$$\implies \frac{f_s}{f_c} = n \left(\frac{600 - x}{x} \right) \quad —$$

Equate (1) & (2) above:

$$\frac{150 x}{3215} = n \left(\frac{600 - x}{x} \right) \quad , \quad \text{use } n =$$

$$x = \frac{-100 \pm \sqrt{(100)^2 - 4(1)(-60,000)}}{2(1)} = 200 \text{ mm}$$

(location of the neutral

$$M_w = C\left(d - \frac{x}{3}\right) \equiv T\left(d - \frac{x}{3}\right)$$

$$226 = T\left(\frac{600}{1000} - \frac{1}{3}\left(\frac{200}{1000}\right)\right)$$

$$\Rightarrow T = C = 424 \text{ kN}$$

$$C = T = \frac{1}{2} f_c \, x \, b \quad \Rightarrow \quad 424 = \frac{1}{2} f_c \left(\frac{200}{1000}\right)\left(\frac{300}{1000}\right)$$

$$\Rightarrow f_c = 14,133 \text{ kN/m}^2 = \boxed{14.13 \text{ MPa}}$$

$$T = f_s A_s \quad \Rightarrow \quad 424 = f_s \left(4 \times \frac{\pi (32)^2}{4}\right) \times 10^{-6}$$

$$\Rightarrow f_s = 131,800 \text{ kN/m}^2 = \boxed{131.8 \text{ MPa}}$$

4.5 Method of Transformed Section:

* The reinforced concrete cross-section is <u>transformed</u> a homogeneous section of one material having the mo of elasticity of concrete, E_c.

* Replace the actual steel area by an equivalent are (imaginary area) in concrete.

* A_s : actual steel area

  A_t : equivalent concrete area (transformed area)

  Total Force in Steel Area = Total Force in Transfor

  $$f_s A_s = f_t A_t$$

  Elongation in Steel Area = Elongation in Transform

  $$\frac{f_s}{E_s} = \frac{f_t}{E_c}$$

  $$\Rightarrow \frac{E_s}{E} = \frac{f_s}{f} \dots \quad \Rightarrow \quad \boxed{f_t = \frac{f_s}{}}$$

4.6 Investigation of Rectangular Sections in Bending
with Tension Reinforcement Only:

* <u>Analysis Problems</u> :

Given: { cross-sectional dimensions and reinforcement ar
         modular ratio
         allowable working stresses

Determine either : 1. compare actual working str
                       with allowable working str

                   2. determine the allowable servi
                      load bending moment the
                      section can carry.

* <u>Note</u> : The <u>neutral axis</u> coincides with the <u>centroidal ax</u>
only in a section under pure bending.

* <u>Three</u> cases are available:

Case 1 : the neutral axis happens to be situated so that t
allowable stresses for both steel and concrete are
reached simultaneously.
⟹ the neutral axis is at the <u>ideal locatic</u>

Case 2 : <u>Steel controls</u> : the neutral axis is nearer
the compression face than the ideal location
only the allowable steel stress can be reach
under the allowable service loading.

Case 3 . <u>Concrete controls</u> : the neutral axis is farth
from the compressive face than the ideal l
only the allowable concrete stress can be rea

* Generally, when $\rho < 75\%$ of ρ_b , concrete contr

* there is no <u>explicit</u> upper limit on ρ in the worki
stress method.
there is an implicit upper limit on ρ when use of

Example 2 :

Given $F_c' = 20$ MPa, allowable $f_s = 135$ MPa, and the ACI Code, determine the allowable service load bending M_w that the rectangular section as shown in the can carry. Locate the elastic neutral axis by the transformed section method; then find M_w using the internal-force and the transformed section.

Solution.

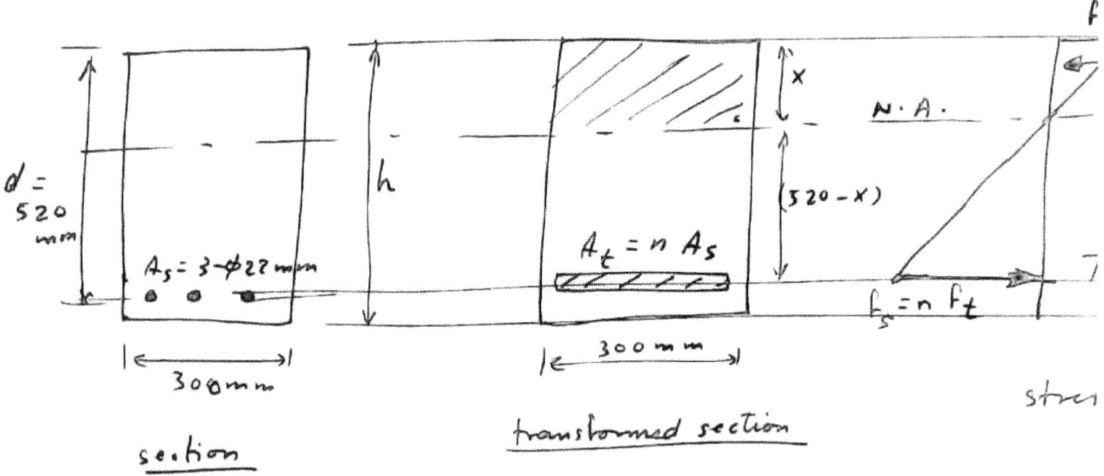

(a) For $f_c' = 20$ MPa $\Rightarrow$ $n = 9$.

$$A_s = 3\left[\frac{\pi (22)^2}{4}\right] = 1140 \text{ mm}^2.$$

$$A_t = n A_s = 9(1140) = 10,264 \text{ mm}^2.$$

Taking moments of areas about N·A·:

$$300 x \left(\frac{x}{2}\right) = 10,264 (520 - x)$$

$$\Rightarrow \quad 150 x^2 + 10,264 x - 5,337,280 = 0$$

$$\Rightarrow x = \frac{-10,264 \pm \sqrt{(10,264)^2 - 4(150)(-5,337,280)}}{2(150)}$$

to the con

(b) $f_{c_{allowable}} = 0.45 f_c' = 0.45(20) = 9$ MPa.

By similar triangles,

$$\frac{9}{157} = \frac{f_t}{520 - 157}$$

$\Longrightarrow f_t = 20.8$ MPa. (if $f_{c_{allowable}}$ is used).

But $f_s = n f_t$

$\therefore f_{s_{allowable}} = n f_{t_{allowable}}$

$135 = 9 f_{t_{allowable}} \Longrightarrow f_{t_{allowable}} = 15$

But $f_t = 20.8$ MPa > 15 MPa. $\Longrightarrow$ $\boxed{Ste}$

$\therefore$ Use $f_t = f_{t_{allowable}} = 15$ MPa.

$\therefore$ By similar triangles:

$$\frac{f_c}{157} = \frac{15}{520 - 157}$$

$\Longrightarrow f_c = 6.49$ MPa < 9 MPa.

$\therefore M_w = C\left(d - \frac{x}{3}\right) \quad or \quad = T\left(d - \frac{x}{3}\right)$

$= T\left(d - \frac{x}{3}\right)$

$= \frac{1}{2} f_c x b \left(d - \frac{x}{3}\right)$

$= \frac{1}{2}(6.49 \times 1000)\left(\frac{157}{1000}\right)\left(\frac{300}{1000}\right)\left[\frac{520}{1000} - \frac{1}{3}\left(\frac{1}{1}\right.\right.$

$= 71.5$ kN·m.

$$\implies \quad M_w = 78.9 \text{ kN} \cdot \text{m}.$$

$$\underset{\text{or}}{=} \quad f_t = \frac{My}{1}$$

$$\implies \quad 15 \times 1000 = \frac{M_w (520-157) \times 10^{-3}}{1739.5 \times 10^{-6}}$$

$$\implies \quad M_w = 78.9 \text{ kN} \cdot \text{m}.$$

Example 3 :

For the rectangular section in Example 2 , dete allowable service load bending moment M_w if the reinforcement is increased from $3-\phi 22$ mm , first $3-\phi 28$ mm, and then to $3-\phi 36$ mm .

Solution :

① $A_s = 3 \left[\frac{\pi (28)^2}{4} \right] = 1847 \text{ mm}^2$.

$$\implies A_t = n A_s = 9(1847) = 16,625 \text{ mm}^2.$$

$$300 x \left(\frac{x}{2} \right) = 16,625 (520 - x)$$

$$150 x^2 + 16,625 x - 16,625 (520) = 0$$

$$x = \frac{-16,625 \pm \sqrt{(16,625)^2 - 4(150)(16,625)(-520)}}{2(150)}$$

$$\implies x = 191 \text{ mm}. \quad [\approx \underline{\text{ideal location}}].$$

$$f_{c\,\text{allowable}} = 0.45 f_c' = 9 \text{ MPa}.$$

$$\frac{9}{191} = \frac{f_t}{520 - 191}$$

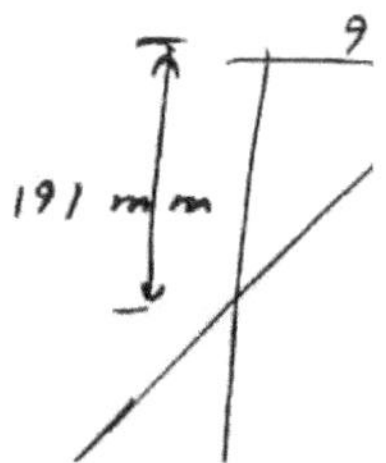

$$\implies f_t = 15.5 \text{ MPa}.$$

$$\frac{f_c}{191} = \frac{15}{520-191}$$

$$\Rightarrow f_c = 8.71 \quad MPa.$$

(<u>Note</u>: this is very close to 9 MPa).

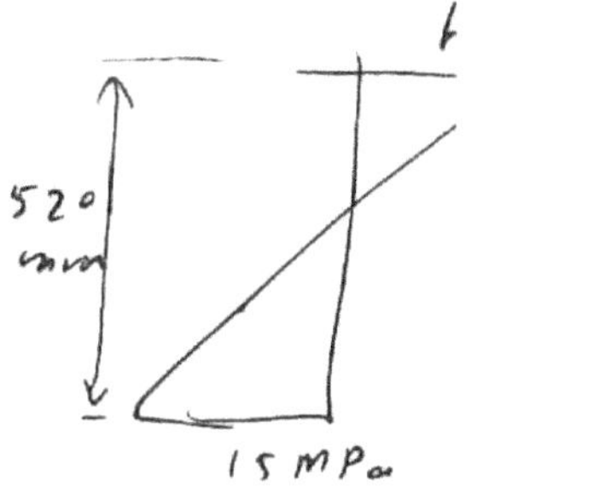

$$M_w = C\left(d - \frac{x}{3}\right)$$

$$= \frac{1}{2} f_c \times b \left(d - \frac{x}{3}\right)$$

$$= \frac{1}{2}(8.71\times1000)\left(\frac{191}{1000}\right)\left(\frac{300}{1000}\right)\left[\frac{520}{1000} - \frac{1}{3}\left(\frac{191}{1000}\right)\right]$$

$$= 113.9 \quad kN \cdot m.$$

---

② $A_s = 3\left[4\frac{(36)^2}{4}\right] = 3,054 \ mm^2.$

$A_t = nA_s = 9(3,054) = 27,483 \ mm^2.$

$$300 \, x\left(\frac{x}{2}\right) = 27,483(520-x)$$

$$150x^2 = 27,483x - 520(27,483)$$

$$\Rightarrow x = \frac{-27,483 \pm \sqrt{(27,483)^2 - 4(150)(-520)(27,483)}}{2(150)}$$

$$\Rightarrow x = 231 \ mm. \quad \left(\begin{array}{l}\text{Farther from the co}\\ \text{face than the ideal}\end{array}\right.$$

$f_{c \ allowable} = 9 \ MPa.$

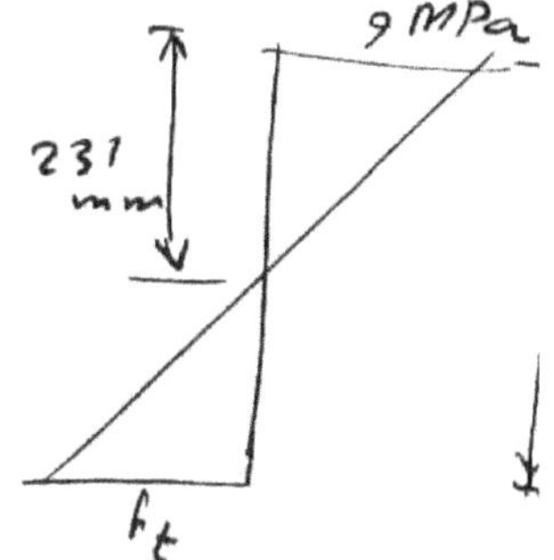

$$\frac{9}{231} = \frac{f_t}{520-231}$$

$$\Rightarrow f_t = 11.3 \quad MPa.$$

$$< 15 \ MPa.$$

$$\therefore \ take \ f_c = 9 \ mPa.$$

$\therefore M_w = C\left(d - \frac{x}{3}\right) \qquad \Longrightarrow \qquad \lfloor \text{Concrete Con}$

<u>Note</u>:

For $A_s = 3-\phi\,36\ mm \implies A_s = 3054\ mm^2$.

$$\rho = \frac{A_s}{bd} = \frac{3054}{(300)(520)} = 0.0196 = 1.96\%.$$

Using the Strength Design Method: ($f_c' = 20\,MPa$, $f_y = 280$

$$\rho_b = \frac{0.85 f_c'\,\beta_1}{f_y}\left(\frac{600}{600+f_y}\right) = \frac{0.85(20)}{280}(0.85)\left(\frac{600}{600+280}\right)$$

$$= 0.0352.$$

$$\rho_{max.} = 0.75\,\rho_b = 0.75\,(0.0352) = 0.0264 = 2.64$$

$\therefore$ In this alternate design method,

$$\rho << \rho_{max}.$$

$\implies$ Steel reinforcement permitted in the working stress design is not as large as the amount permitted in the strength design.

4.7 Design of Rectangular Sections in Bending with Tension Reinforcement Only:

Design Problems in Alternate Design Method:

<u>Given</u>: 1. bending moment, M
2. modular ratio, n.
3. allowable working stresses.

<u>Determine</u>: b, d, A_s.

* Assume the neutral axis is at the ideal location:

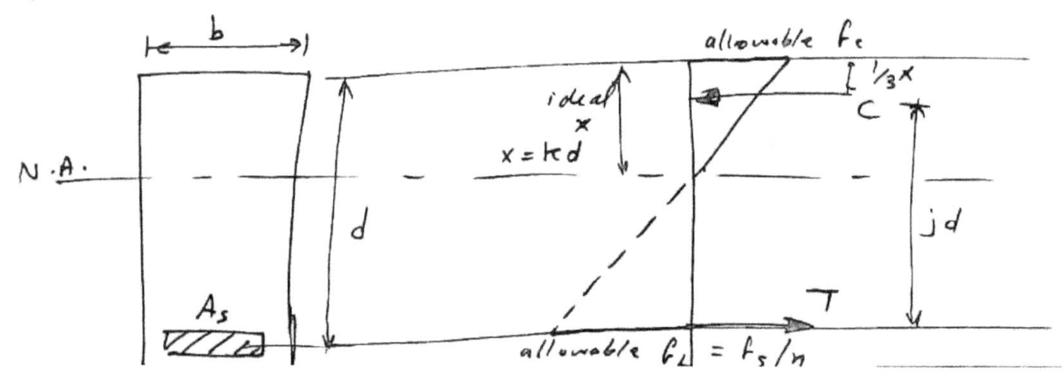

Let $x = kd$ (ideal) and moment arm $= jd$ (ideal).

$$\frac{kd}{\text{allowable } f_c} = \frac{d - kd}{\text{allowable } f_t}$$

$$(\text{allowable } f_t)\, kd = (\text{allowable } f_c)\, d - (\text{allowable } f_c)$$

$$(\text{allowable } f_s/n + \text{allowable } f_c)\, kd = (\text{allowable } f_c)\, d$$

$$\Longrightarrow \boxed{k = \frac{\text{allowable } f_c}{\text{allowable } f_s/n + \text{allowable } f_c}} \Longrightarrow \text{1st de}$$

ideal moment arm $jd = d - \frac{x}{3} = d - \frac{kd}{3} = (1 - \frac{k}{3}$

$$\Longrightarrow \boxed{j = 1 - \frac{k}{3}} \Longrightarrow \text{2nd design constant}$$

$$C = T$$

$$(\tfrac{1}{2}\,\text{allowable } f_c)\, b x = (\text{allowable } f_s)\, A_s$$

$$\tfrac{1}{2}(\text{allowable } f_c)\, b\, kd = (\text{allowable } f_s)\, \rho\, bd$$

$$\Longrightarrow \boxed{\rho = \frac{k}{2}\frac{\text{allowable } f_c}{\text{allowable } f_s}} \quad \text{at the ideal} \Longrightarrow \text{3rd design}$$

$$M_w = C \cdot jd = \tfrac{1}{2}(\text{allowable } f_c)(b\,kd)(jd)$$

$$= \tfrac{1}{2}(\text{allowable } f_c)\, jk\, bd^2 = Rbd$$

where $\boxed{R = \tfrac{1}{2}(\text{allowable } f_c)\, jk} \Longrightarrow \text{4th design}$

Also, $M_w = T \cdot jd = (\text{allowable } f_s)\, A_s \cdot jd$

Generally, the reinforcement ratio at the ideal location is half that permitted by the strength method.

Procedure for Determining the Theoretical Values of b, d, and A_s for placing the elastic neutral axis at the ideal location:

(1) Find the required value of bd^2 and M_w/R.

(2) Assume a value of b and determine d (select b and h and check weight).

(3) Determine A_s from $\rho b d$ and check its value from

$$\frac{M_w}{(\text{allowable } f_s)(jd)}.$$

Example 4:

Design the cross-section for a rectangular beam to carry a uniform live load of 28 kN/m and a uniform dead load of 15 kN/m (not including beam weight) on a simply-supported span of 10 m. Use the ACI Code with $f_c' = 28$ MPa and $f_y = 415$ MPa (Grade 60 steel).

Solution:

(a) Determine the four design constants to place the elastic neutral axis at the ideal location:

$$\text{allowable } f_c = 0.45 f_c' = (0.45)(28) = 12.6 \text{ MPa}.$$
$$(\text{ACI}/\text{Appendix}/A\,3.1).$$

$$\text{allowable } f_s = 170 \text{ MPa}. \quad (\text{ACI}/\text{Appendix}/A\,3.2).$$
$$(\text{for Grade 60 steel}).$$

For $f_c' = 28$ MPa, $n = 8$. (take the higher value).

$$k = \frac{\text{allowable } f_c}{\text{allowable } f_s/n + \text{allowable } f_c} = \frac{12.6}{\frac{170}{8} + 12.6}$$
$$= 0.372.$$

$$j = 1 - \frac{k}{3} = 1 - \frac{0.372}{3} = 0.876.$$

$$R = \frac{1}{2}\left(\text{allowable } f_c\right)jk = \frac{1}{2}(12.6)(0.876)(0.372) = 2.05 \text{ MPa}.$$

$$\rho = \frac{k}{2} \frac{\text{allowable } f_c}{\text{allowable } f_s} = \left(\frac{0.372}{2}\right)\left(\frac{12.6}{170}\right) = 0.0138 = 1.38\%.$$

(b) Determine preliminary size: First estimate beam weight:

Take $b = 450$ mm and $h = 800$ mm.

$$w = 25 \times \frac{450}{1000} \times \frac{800}{1000} = 9 \text{ kN/m}.$$

$$w_d = 9 + 15 = 24 \text{ kN/m}.$$

Note: In this method, service loads are **not** multiplied by load factors.

$$M_{w(d)} = \frac{1}{8} w_d L^2 = \frac{1}{8}(24)(10)^2 = 300 \text{ kN·m}.$$

$$M_{w(\ell)} = \frac{1}{8} w_\ell L^2 = \frac{1}{8}(28)(10)^2 = 350 \text{ kN·m}.$$

$$\therefore \quad M_w = M_{w(d)} + M_{w(\ell)} = 300 + 350 = 650 \text{ kN/m}.$$

To place the elastic N.A. at the ideal location,

$$\text{required } bd^2 = \frac{M_w}{R} = \frac{650}{2.05 \times 1000} = 0.317 \text{ m}^3$$
$$= 317 \times 10^6 \text{ mm}^3.$$

For $b = 450$ mm,

$$\text{required } d = \sqrt{\frac{317 \times 10^6}{450}} = 839 \text{ mm}.$$

Now, determine overall depth h:

Assume two layers of equal-sized bars:

$$h = d + 40 + 10 + 25 + \frac{1}{2}(25)$$
$$\therefore h = d + 87.5 \quad (\text{for two layers}).$$

For two-layers of bars, usually, we have:

$$h = d + (90 \rightarrow 100 \text{ mm}).$$

thus, $h = 839 + 87.5 = 926.5$ mm.

$\Rightarrow$ Use a whole number for h.

$\therefore$ Take $h = 950$ mm.

$\Rightarrow$ Try a beam with $b = 450$ mm, $h = 950$ mm.

(c) Check weight and revise step (b) if necessary:

$$w = 25 \times \frac{450}{1000} \times \frac{950}{1000} = 10.69 \text{ kN/m}.$$

$$w_d = 10.69 + 15 = 25.69 \text{ kN/m}.$$

$$\therefore M_{w(d)} = \frac{1}{8}(25.69)(10)^2 = 321.1 \text{ kN·m}.$$

$$\Rightarrow M_w = M_{w(d)} + M_{w(e)} = 321.1 + 350 = 671 \text{ kN·m}.$$

$$\therefore \text{required } bd^2 = \frac{M_w}{R} = \frac{671}{2.05 \times 1000} = 0.3273 \text{ m}^3$$
$$= 327.3 \times 10^6 \text{ mm}^3.$$

$$\therefore d = \sqrt{\frac{327.3 \times 10^6}{450}} = 853 \text{ mm}.$$

$$\therefore \text{required } h = d + (\approx 100 \text{ mm})$$
$$= 853 + 100$$
$$= 953 \text{ mm} \approx 950 \text{ mm}.$$

$\therefore$ the chosen beam (450×950) is acceptable.

(d) Determine required A_s:

Use $\phi 28$ mm bars.

$$\therefore d = h - 40 - 10 - 28 - \frac{1}{2}(25)$$
$$= h - 90$$
$$\Rightarrow d = 950 - 90 = 860 \text{ mm}.$$

$$\therefore \text{ required } A_s = \frac{M_w}{(\text{allowable } f_s)\, jd} = \frac{671}{(170 \times 1000)(0.876)\left(\frac{860}{1000}\right)}$$

$$= 0.005239 \ m^2$$

$$= 5239 \ mm^2.$$

Using the tables, Try $8 - \phi 30\,mm \Rightarrow A_s = 5654 \ mm^2.$

New $d = h - 40 - 10 - 30 - \frac{1}{2}(23) = h - 92.5$

$$= 857.5 \approx 858\,mm.$$

$$\therefore \text{ required } A_s = \frac{671}{(170 \times 1000)(0.876)\left(\frac{858}{1000}\right)} \times 10^6$$

$$= 5251 \ mm^2.$$

$$\text{used } A_s = 5654 \ mm^2 \qquad \underline{\underline{o.k.}}$$

(d) Check the section:

Take moments of areas about N.A.

$$A_t = nA_s = 8(5654) = 45,232 \ mm^2.$$

$$(bx)\left(\frac{x}{2}\right) = A_t (d - x)$$

$$(450)\frac{x^2}{2} = 45,232 \,(858 - x)$$

$$225x^2 + 45,232x - 45,232(858) = 0$$

$$x = \frac{-45,232 \pm \sqrt{(45,232)^2 - 4(225)(45,232)(-858)}}{2(225)}$$

$$\Rightarrow x = 327 \ mm.$$

$$M_w = C\left(d - \frac{x}{3}\right) \qquad \text{or} \qquad T\left(d - \frac{x}{3}\right)$$

$$671 = \tfrac{1}{2} f_c bx \left(d - \frac{x}{3}\right)$$

$$671 = \tfrac{1}{2} f_c \left(\frac{450}{1000}\right)\left(\frac{327}{1000}\right)\left[\frac{858}{1000} - \tfrac{1}{3}\left(\frac{327}{1000}\right)\right]$$

$$\Rightarrow \; f_c = 12176 \; kN/m^2 \quad = \; 12.18 \; MPa$$
$$< \; \text{allowable} \; f_c = 12.6 \; MPa.$$

$$M_w = T \left(d - \frac{x}{3} \right)$$
$$671 = f_s A_s \left(d - \frac{x}{3} \right)$$
$$671 = f_s \left(5654 \times 10^{-6} \right) \left[\frac{858}{1000} - \frac{1}{3} \left(\frac{327}{1000} \right) \right]$$
$$\Rightarrow \; f_s = 158447 \; kN/m^2$$
$$= 158 \; MPa$$
$$< \; \text{allowable} \; f_s = 170 \; MPa.$$

<u>Note</u>: The actual stresses are very close to the allowable stresses.

$\Rightarrow$ N.A. is $\approx$ at the ideal location.

<u>Check</u>: If bars will fit in beam width.

Provide Design Sketch:

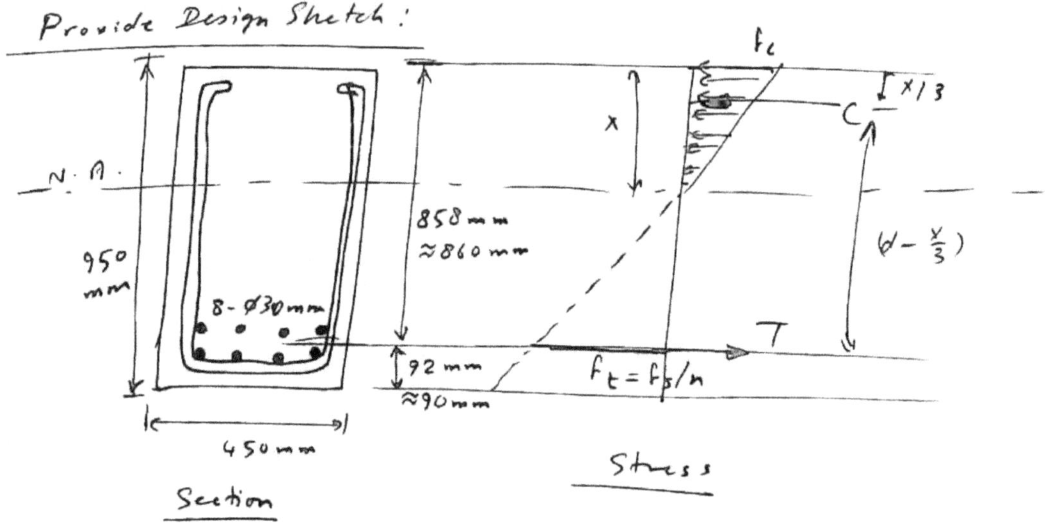

$$\text{clear distance between bars} = \frac{450 - 2(40) - 2(10) - 4(30)}{3} = 77 \; mm$$
$$> 30 \; mm.$$
$$\underline{o.k}$$

Example 5:

For the service load moment $M_w = 671$ kN·m (final M_w for Example 4), determine the required A_s if a section 450×850 nd were to be used. Use $f_c' = 28$ MPa, $f_y = 415$ MPa, and the ACI Code. (The value of M_w could be revised due to the reduction in weight of beam, but the purpose here is to illustrate the increase in required A_s for the same M_w).

Solution:

(a) Design constants are unchanged from Example 4:

$$k = 0.372 \quad , \quad j = 0.876 \quad , \quad R = 2.05 \text{ MPa} .$$

(b) Determine the required position of the elastic neutral axis relative to the ideal location.

For two layers of steel, $d = h - 100 = 850 - 100 = 750$ mm.

$$\therefore \quad \text{required } R = \frac{M_w}{bd^2} = \frac{671/1000}{\left(\frac{450}{1000}\right)\left(\frac{750}{1000}\right)^2} = 2.65 \text{ MPa} .$$

Since required $R = 2.65$ > ideal $R = 2.05$ MPa

$\Rightarrow$ location of elastic N.A. is larger than the ideal location.

$\Rightarrow$ $\boxed{\text{Concrete Controls in this Example}}$, $f_c \equiv$ allowable f_c.

(allowable)

$= 12.6$ MPa.

(c)
$$M_w = C\left(d - \frac{x}{3}\right) = \frac{1}{2} f_c (bx)\left(d - \frac{x}{3}\right)$$

$$671 = \frac{1}{2}(12.6 \times 1000)\left(\frac{450}{1000}\right) x \left(\frac{750}{1000} - \frac{1}{3} x\right)$$

$$0.23668 = x(0.75 - 0.3333x)$$

$$0.3333 x^2 - 0.75 x + 0.23668 = 0$$

$$x^2 - 2.25 x + 0.71 = 0$$

$$x = \frac{+2.25 \pm \sqrt{(2.25)^2 - 4(1)(0.71)}}{2(1)} = 0.380 \text{ m}$$

$$= 380 \text{ mm} .$$
$$> 327 \text{ mm (ideal location)}$$

$$\frac{12.6}{380} = \frac{f_t}{750 - 380}$$

$$\implies f_t = 12.27 \ MPa.$$

$$\therefore \ f_s = n f_t = 8(12.27)$$
$$= 98.15 \ MPa.$$

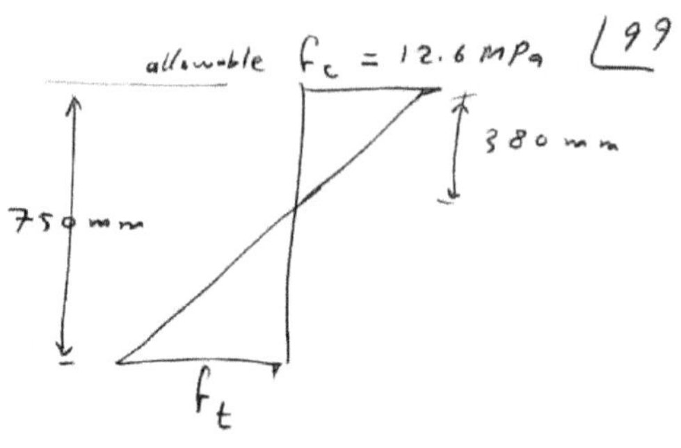

Note: (This is very much less than allowable $f_s = 170 \ MPa$).

$\implies$ the strength of the section is <u>not</u> reached using this design method.

$$M_w = T\left(d - \tfrac{x}{3}\right) = f_s A_s \left(d - \tfrac{x}{3}\right)$$

$$671 = (98.15 \times 1000) A_s \left[\frac{750}{1000} - \tfrac{1}{3}\left(\frac{380}{1000}\right) \right]$$

$$\implies A_s = 0.01097 \ m^2$$
$$= 10968 \ mm^2$$

Note: required $A_s = 10968 \ mm^2 > A_{s(ideal)} = 5251 \ mm^2$.

$$\implies \boxed{Over\text{-}reinforced \ Beam}$$

Note: the major <u>drawback</u> of this design method is that a very large area of reinforcement is used while the steel stress f_s is very small.

Alternative: use compression reinforcement.

Example 6 :

For the service load moment $M_w = 671 \ kN \cdot m$ (final M_w for Example 4), determine the required A_s if a section $450 \times 1000 \ mm$ were to be used. Use $f_c' = 28 \ MPa$, and $f_y = 415 \ MPa$, and the ACI Code. (The value of M_w should be revised due to increase in weight of the beam, but the purpose here is to illustrate the decrease in required A_s for the same M_w.

Solution :

(a) Design constants are unchanged from Example 4.

$$K = 0.372 \quad , \quad j = 0.876 \quad , \quad R = 2.05 \ \text{MPa}.$$

(b) Determine the required position of the elastic neutral axis relative to the ideal location.

For two layers of steel, $d = h - 100 = 1000 - 100 = 900 \ mm$.

$$\text{required} \ R = \frac{M_w}{b d^2} = \frac{671/1000}{\left(\frac{450}{1000}\right)\left(\frac{900}{1000}\right)^2} = 1.84 \ \text{MPa}.$$

Since required $R = 1.84 \ MPa <$ ideal $R = 2.05 \ MPa$

$\Rightarrow$ location of elastic neutral axis is smaller than the ideal location.

$$\Rightarrow \boxed{\text{Steel Controls in this Example}}$$

$$\Rightarrow \ f_s = \text{allowable} \ f_s = 170 \ MPa.$$

(c) $f_t = f_s/n = \dfrac{170}{8} = 21.25 \ MPa.$

$$\frac{f_a}{x} = \frac{21.25}{900 - x}$$

$$f_a = \frac{21.25 \, x}{900 - x}$$

$$M_w = C\left(d - \frac{x}{3}\right) = \frac{1}{2} f_c \, bx \left(d - \frac{x}{3}\right)$$

$$671 = \frac{1}{2}\left(\frac{21.25 x (1000)}{0.900 - x}\right)\left(\frac{450}{1000}\right) x \left(\frac{900}{1000} - \frac{1}{3} x\right)$$

$$671 = \frac{1}{2}\left(\frac{21.25 \, x}{0.900 - x}\right)(450 x)\left(0.9 - \frac{1}{3} x\right)$$

$$1342(0.900 - x) = (21.25 x)(450 x)(0.9 - 0.3333 x)$$

$$1207.8 - 1342 x = 8606.25 x^2 - 3187.5 x^3$$

$$x^3 - 2.7 x^2 - 0.421 x + 0.3789 = 0$$

Solve the cubic equation by numerical methods.

$$x = 320 \ mm. \quad (\text{see next page}).$$
$$< 327 \ mm \ (\text{ideal location}).$$

Solving Nonlinear Equations by the Newton-Raphson Method:

* To solve the equation $f(x) = 0$:

1. Find the derivative $f'(x)$.

2. Start with an initial estimate x_0.

3. Iterate $x_{n+1} = x_n - \dfrac{f(x)}{f'(x)}$

4. Stop when the error is small.

Example:

Solve the equation $x^3 - 2.7x^2 - 0.421x + 0.3789 = 0$

Let $f(x) = x^3 - 2.7x^2 - 0.421x + 0.3789$

$f'(x) = 3x^2 - 5.4x - 0.421$

$\therefore \quad x_{n+1} = x_n - \dfrac{f(x_n)}{f'(x_n)}$

$\implies x_{n+1} = x_n - \dfrac{x^3 - 2.7x^2 - 0.421x + 0.3789}{3x^2 - 5.4x - 0.421}$

Start with $x_0 = 0.300$ m.

$\implies x_1 = 0.321$ m.

$\implies x_2 = 0.320$ m.

$$\implies \boxed{x = 320\,mm.}$$

$$\therefore \; f_c = \frac{21.25\,x}{900 - x} = \frac{(21.25 \times 1000)(320/1000)}{\left(\frac{900}{1000} - \frac{320}{1000}\right)}$$

$$\Rightarrow \quad f_c = 11{,}724 \quad kN/m^2$$

$$= 11.724 \quad MPa \leqslant allowable \; f_c = 17.6 \; MPa.$$

$$M_w = T\left(d - \frac{x}{3}\right)$$

$$\therefore \quad M_w = f_s A_s \left(d - \frac{x}{3}\right)$$

$$671 = (170 \times 1000)\, A_s \left[\frac{900}{1000} - \frac{1}{3}\left(\frac{320}{1000}\right)\right]$$

$$\Rightarrow \quad A_s = 0.004975 \; m^2$$

$$= 4975 \; mm^2.$$

Note: required $A_s = 4975 \; mm^2 < A_{s(ideal)} = 5251 \; mm^2$

$$\Rightarrow \quad \boxed{Under\text{-}reinforced \; Beam}$$

Note: Instead of solving the cubic equation for x,
we compare the moment arms:

$$\left(d - \frac{x}{3}\right) = 900 - \frac{1}{3}(320) = 793 \; mm.$$

$$jd = (0.876)(900) = 788 \; mm.$$

The actual moment arm 793 mm is only 5 mm larger than the ideal jd.

$\Rightarrow$ An approximate value of A_s for "under-reinforced sections"

"only" can be obtained by:

$$A_{s\,(approximate)} = \frac{M_w}{(allowable \; f_s)(j)(actual \; d)}$$

$$= \frac{671}{(170 \times 1000)(0.876)(0.900)} = 0.005006 \; m^2$$

$$= 5006 \; m^2$$

very close to the correct value of 4975 mm² and is on the safe side.

4.8 Design of Rectangular Sections in Bending with Both Tension and Compression Reinforcement:

* Use the method of superposition.

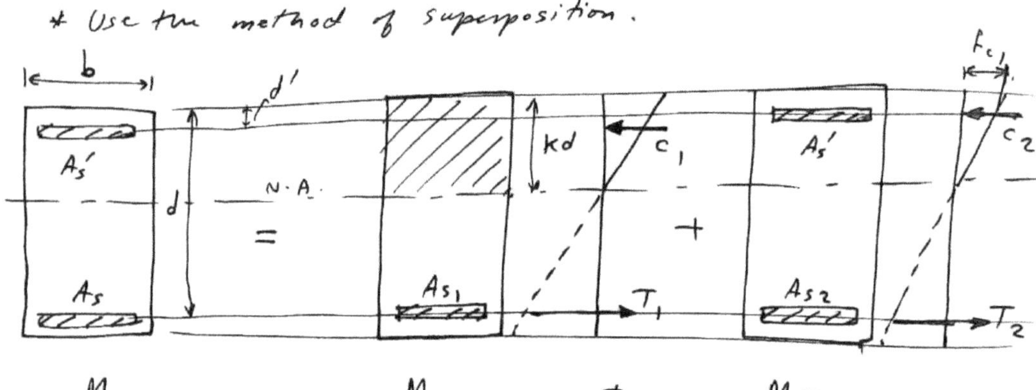

$$M_w \qquad = \qquad M_{w_1} \qquad + \qquad M_{w_2}$$

Steps :

1. Compute $M_{w_1} = R b d^2$.

2. Compute $A_{s_1} = \rho b d$

   Check $A_{s_1} = \dfrac{M_{w_1}}{(\text{allowable } f_s)(j d)}$

3. Compute $M_{w_2} = M_w - M_{w_1}$

4. Compute $C_2 = T_2$ from $M_{w_2} = C_2 (d - d') = T_2 (d - d')$

5. Compute A_{s_2} from $T_2 = (\text{allowable } f_s) A_{s_2}$

6. Compute $A_s = A_{s_1} + A_{s_2}$

7. Compute f_{c_1} from similar triangles :

$$\frac{f_{c_1}}{kd - d'} = \frac{\text{allowable } f_c}{kd}$$

$$\implies f_{c_1} = \left(\frac{kd - d'}{kd}\right)(\text{allowable } f_c)$$

8. Compare $2 n f_{c_1}$ with allowable f_s.

9. Compute $A_s' = \dfrac{C_2}{(2n-1) f_{c_1}}$, if $2 n f_{c_1} \leq$ allowable f_c

$$A_s' = \frac{C_2}{(\text{allowable } f_s - f_{c_1})} , \text{ if } 2 n f_{c_1} > \text{allowable } f_s$$

See ACI - Code / Section / Appendix A5.5.

4.9 Investigation of Rectangular Sections in Bending with Both Tension and Compression Reinforcement:

* Transformed compression steel area $A_t' = (2n-1) A_s'$.
  (when computing stresses).
* Stress in compression steel is either $2nf_{c_1}$ or allowable f_c, whichever is smaller.
* Transformed compression steel area $A_t' = n A_s'$ (when calculating deflection), because creep and shrinkage are accounted for in a different way (see Chapter 14).

$$\frac{f_c}{x} = \frac{f_{c_1}}{x - d'}$$

$$\Rightarrow \quad f_{c_1} = f_c \left(\frac{x - d'}{x} \right)$$

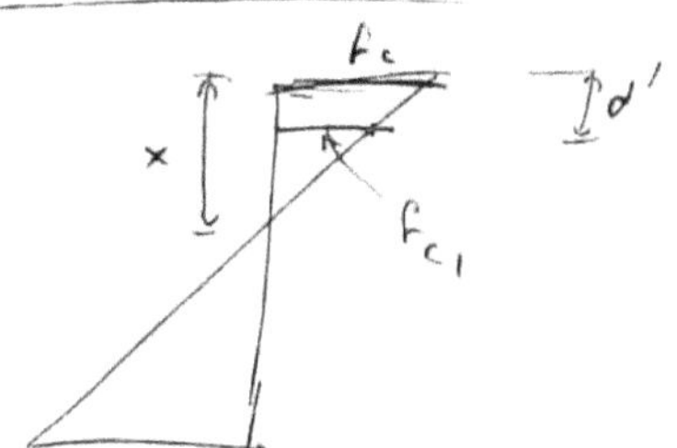

$$\Rightarrow \quad \text{Equilibrium} \quad C_c + C_s = T$$

$$\tfrac{1}{2} f_c \, bx + \left(\underbrace{2nf_{c_1} - f_{c_1}}_{\substack{\text{or} \\ \text{allowable } f_s - f_{c_1}}} \right) A_s' = f_c \left(\frac{x - d'}{x} \right) A_s$$

$$(2n-1) f_{c_1}$$

solve the above equation for x.

Note: No numerical examples (analysis or design) of doubly-reinforced sections are given by the working stress method.

4.10 Serviceability — Deflections:

* Excessive deflection may make the structure unserviceable.

* Deflection computations under service load conditions are necessary, for flexural members that support or are attached to partitions and other construction likely to be damaged by large deflections.

* For members not supporting or attached to partitions or other construction likely to be damaged by large deflections, ACI Table 9.5(a) provides minimum thickness

for beams and one-way slabs unless deflections are computed.

* When the minimum thickness requirement of ACI—Table 9.5(a) is not satisfied, <u>or</u> if excessive deflection may cause a problem, deflection must be computed and must satisfy the limits of ACI—Table 9.5(b).

* Since the applied (service) load moment usually far exceeds the moment that causes the tension concrete to crack (e.g. cracking moment), the cracked section should be used.

* Deflection calculations generally use the formula:

$$\delta = \beta_a \frac{M_w L^2}{E_c I_e}$$

Example 7:

Compute the moment of inertia I_{cr} of the transformed cracked section to be used in "deflection calculations" for the doubly-reinforced section shown in the figure, using $f_c' = 28\ MPa$ $(n = 8)$.

Solution:

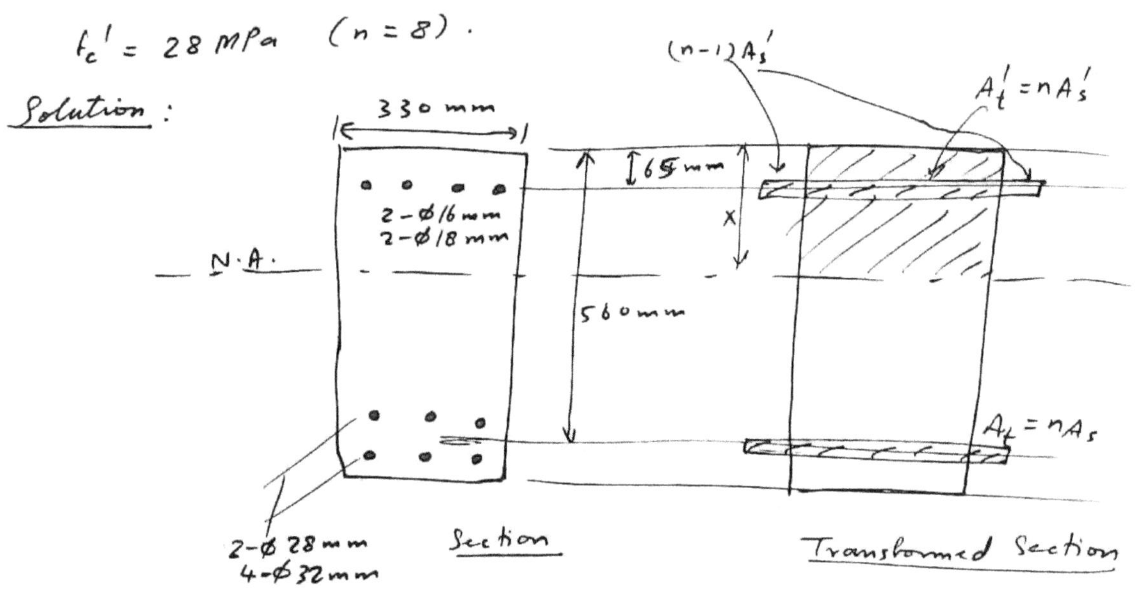

$$A_s = 2\left[\frac{\pi (28)^2}{4}\right] + 4\left[\frac{\pi (32)^2}{4}\right] = 4448\ mm^2.$$

$$A_s' = 2\left[\frac{\pi (16)^2}{4}\right] + 2\left[\frac{\pi (18)^2}{4}\right] = 911\ mm^2.$$

$$A_t = n A_s = 8(4448) = 35584 \ mm^2.$$

$$A_t' = (n-1)A_s' = (8-1)(911) = 6377 \ mm^2.$$

Take moments about N.A.:

$$(330x)\frac{x}{2} + (6377)(x-65) = 35584(560-x)$$

$$\Rightarrow \quad 165x^2 + 41961x - 19512535 = 0$$

$$\therefore \quad x = \frac{-41961 \pm \sqrt{(41961)^2 - 4(165)(-19512535)}}{2(165)} = 217 \ mm.$$

$$I_{cr} = \tfrac{1}{3} b x^3 + (n-1)A_s'(x-d')^2 + nA_s(d-x)^2$$

$$= \tfrac{1}{3}(330)(217)^3 + (7)(911)(217-65)^2 + (8)(4448)(560-217)^2$$

$$= 5458 \times 10^6 \ mm^4.$$

* Deflection calculations are given in Chapter 14.

4.11 Serviceability — Flexural Crack Control for Beams and One-Way Slabs:

* <u>Cracking</u> in concrete is the result of:

(1) volumetric change, including that due to drying shrinkage, creep under sustained load, thermal stresses, and chemical incompatibility of concrete components.

(2) internal or external direct stress due to continuity, reversible load, long-time deflection, or differential movement in structures.

(3) flexural stress due to bending.

* As larger cracks form, the exposure to a corrosive environment may be detrimental to the steel. Factors such as humidity, salt air, alternate wetting and drying, or freezing and thawing may accelerate corrosion and contribute to concrete deterioration in the vicinity of large width cracks.

* Although cracking cannot be eliminated, it is generally more desirable to have many fine hairline cracks than a few wide cracks.

⟹ <u>Crack control</u> is a matter of controlling the distribution and size of cracks, rather than eliminating them.

* To control cracking, it is better to use several smaller bars at moderate spacing than larger bars of equivalent area.

* See ACI Code, Section 10.6 for distribution of flexural reinforcement which contains crack control provisions for beams and one-way slabs.

* Crack control is important in the following two cases.

   (1) when $f_y > 280$ MPa.

   (2) f exceeds f of the working stress method.

* For the strength design method, limit f to $\frac{1}{2} f_{max}$

   $= 0.375 f_b$ in order to control deflections and cracks.

---

* Crack width is proportional to steel stress.

* ACI Code / Section 10.6.4 (Gergely – Lutz, 1968).

$$w = C \beta_h f_s \sqrt[3]{d_c A}$$

when

w : crack width at the tension face of the beam. (in mm).

$\beta_h = \dfrac{h_2}{h_1}$.

f_s : service load stress in steel (in MPa).

d_c : concrete cover (see the figure).

$A = \dfrac{A_e}{m}$, m = number of bars. (in mm^2).

For different bar sizes, $m = \dfrac{A_s}{\text{Area of largest bar}}$

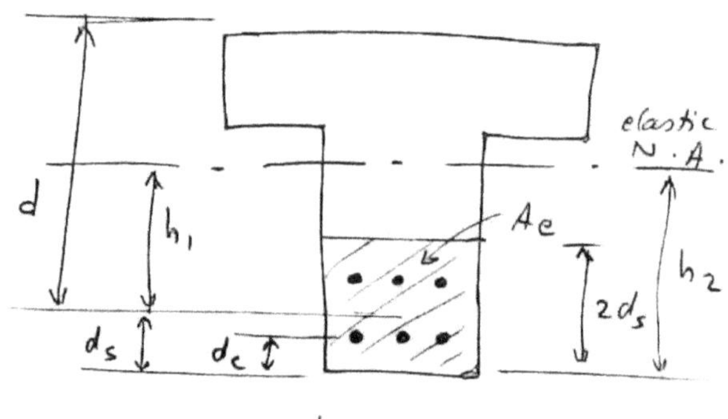

C = an experimental constant.

$$C = 11.0 \times 10^{-6} \ mm^2/N \qquad (SI \ Units).$$
$$C = 76 \times 10^{-6} \ in^2/k \qquad (English \ Units).$$

---

* The above equation gives the most probable maximum crack width on the bottom face of a beam.

* Divide the above equation by $C\beta_h$:

$$w = C\beta_h f_s \sqrt[3]{d_c A}$$

$$\Rightarrow \qquad \frac{w}{C\beta_h} = f_s \sqrt[3]{d_c A}$$

Define the quantity $z = \dfrac{w}{C\beta_h}$

$$\Rightarrow \qquad \boxed{z = f_s \sqrt[3]{d_c A}}$$

Take $\beta_h = \dfrac{h_2}{h_1} \approx 1.2$.

* Crack widths have been limited to $0.41 \ mm$ and $0.33 \ mm$ for interior and exterior exposure, respectively.

$$z = \frac{w}{C\beta_h}$$

(1) For interior exposure, $z_{limit} = \dfrac{0.41}{(11.0 \times 10^{-6})(1.2)} = 31,061 \ N/mm$

$$= \frac{31,061 \times 10^{-6}}{1 \times 10^{-3}}$$

$$= 31 \ MN/m$$

You can use $z_{limit} = 30 \ MN/m$ (interior).

(2) For exterior exposure, $z_{limit} = \dfrac{0.33}{(11.0 \times 10^{-6})(1.2)} = 25,000 \ N/mm$

$$= 25 \ MN/m \quad (exterior).$$

* The above two limits on z are specified by the ACI Code/Section 10.6.4, when $f_y > 280 \ MPa$.

* The <u>Gergely–Lutz</u> formula applies also to one-way slabs, but with $\beta_h = \dfrac{h_2}{h_1} = 1.35$ (instead of 1.2).

$$\therefore \quad z_{limit} = \frac{0.41}{(11 \times 10^{-6})(1.35)} \times 10^{-3} = 27.6 \quad MN/m \quad (interior).$$

$$z_{limit} = \frac{0.33 \times 10^{-3}}{(11 \times 10^{-6})(1.35)} = 22.2 \quad MN/m \quad (exterior).$$

* For two-way slabs, the above limits do not apply.

---

* There is no need to perform a working stress analysis in order to determine f_s needed by the formula.

* ACI Code / Section 10.6.4 permits taking $f_s = 0.6\, f_y$ for crack control calculations. This is very conservative since it will overestimate the value of f_s.

Example 8 :

Check the crack control provisions of the ACI Code against the cross-section selected in Example 4. The selected beam has $b = 450$ mm, $h = 950$ mm, $8-\phi30$ mm bars in two layers, $\phi10$ mm stirrups, 40-mm clear cover at bottom, and 25 mm clear between layers. Take $f_c' = 28$ MPa and $f_y = 415$ MPa (Grade 60 steel).

Solution :

$$z = f_s \sqrt[3]{d_c\, A}$$

Here, $f_s = 158$ MPa (computed in Example 4).

d_c = distance to centroid of bottom layer

$$= 40 + 10 + \tfrac{1}{2}(30)$$

$$= 65 \text{ mm}.$$

d_s = distance to centroid of A_s

$$= 40 + 10 + 30 + \tfrac{1}{2}(25)$$

$$= 92.5 \text{ mm}.$$

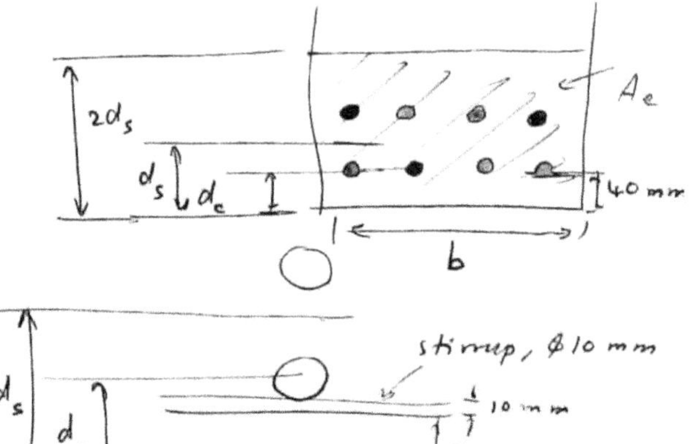

A_e = shaded area $= b(2d_s)$

$$= (450)(2)(92.5) = 83,250 \text{ mm}^2.$$

$m = 8$ (# of bars).

$$A = \frac{A_e}{m} = \frac{83,250}{8} = 10406.25 \text{ mm}^2.$$

$$\therefore \; Z = f_s \sqrt[3]{d_c A} = 158 \sqrt[3]{\left(\frac{65}{1000}\right)\left(\frac{10406.25}{10^6}\right)}$$

$$= 13.9 \text{ MN/m}.$$

Since Z does not exceed the 30 MN/m or 25 MN/m allowed, the selected cross-section is acceptable (satisfies ACI Code crack control provisions).

Example 9 :

Estimate the crack control situation of Example 8 if $2-\phi 58$ mm bars had been used instead of $8-\phi 30$ mm .

Solution :

For one layer of reinforcement,

$$d_c = d_s = 40 + 10 + \frac{1}{2}(58)$$
$$= 79 \text{ mm}.$$

Now, compute f_s :

Take moments of areas about N.A :

$$(450x)\frac{x}{2} = (n A_s)(d - x)$$

$$225 x^2 = 8\left[\pi \frac{(58)^2}{4}(2)\right](860 - x)$$

$$225 x^2 = 42273 (860 - x)$$

$$x^2 = 187.88 (860 - x)$$

$$x^2 + 187.88x - 187.88(860) = 0$$

$$\Rightarrow \quad x = \frac{-187.88 \pm \sqrt{(187.88)^2 - 4(-187.88)(860)}}{2(1)} = 319 \text{ mm}.$$

$$M_w = T\left(d - \frac{x}{3}\right) = f_s A_s \left(d - \frac{x}{3}\right)$$

$$671 = f_s \left[2\,\frac{\pi (58)^2}{4}\right] \times 10^{-6} \left[\frac{860}{1000} - \frac{1}{3}\left(\frac{319}{1000}\right)\right]$$

$$\Rightarrow \quad f_s = 168,487 \ \ kN/m^2$$

$$= 168 \ \ MPa .$$

[$\underline{Note}$: if the elastic analysis is $\underline{not}$ made, you can take a conservative value of $f_s = 0.6\,f_y = 249\,MPa$].

$$A_e = b(2d_s) = (450)(2)(79) = 71,100 \ \ mm^2 .$$

$$m = 2$$

$$A = \frac{A_e}{m} = \frac{71,100}{2} = 35,550 \ \ mm^2 .$$

$$z = f_s \sqrt[3]{d_c A} = (168)\sqrt[3]{(79)(35,550)} \quad \begin{array}{l} = 23,703 \ N/mm \\ = 23.7 \ MN/m . \\ < 25 \ MN/m . \end{array}$$

$\underline{Note}$: If $f_s = 0.6\,f_y = 249\,MPa$ is used,

$$then \quad z = \left(\frac{249}{1000}\right)\sqrt[3]{(79)(35,550)} \quad \begin{array}{l} = 35.1 \ MN/m \\ > 25 \ MN/m . \end{array}$$

therefore, you should use the elastic analysis.

---

* when using bundled bars, use $m = 1.4$, regardless of the number of bars in the bundle.

4.12 Serviceability — Side Face Crack Control for Large Beams:

* On $\underline{deep\ beams}$, the maximum crack width may occur along the side faces of the beams between the neutral axis and the main tension reinforcement.

* Unacceptably wide $\underline{side\ face\ cracks}$ have been observed that are three times as wide as the crack width at the main tension reinforcement.

* ACI Code/Section 10.6.7 has provisions requiring side face $\underline{(skin)\ reinforcement}$.

* When the depth of the web exceeds 1 m, longitudinal skin reinforcement must be provided.

* The _skin reinforcement_ A_{sk} (mm² per meter of height, on each side) shall be :

$$A_{sk} \geq 1.0\,(d - 750)$$

* The maximum spacing of the skin reinforcement may not exceed the lesser of $d/6$ and 300 mm, and the total area in both faces as skin reinforcement need not exceed one-half of the required flexural tension reinforcement.

* Skin reinforcement must be placed within a distance of $d/2$ from the main tension reinforcement.

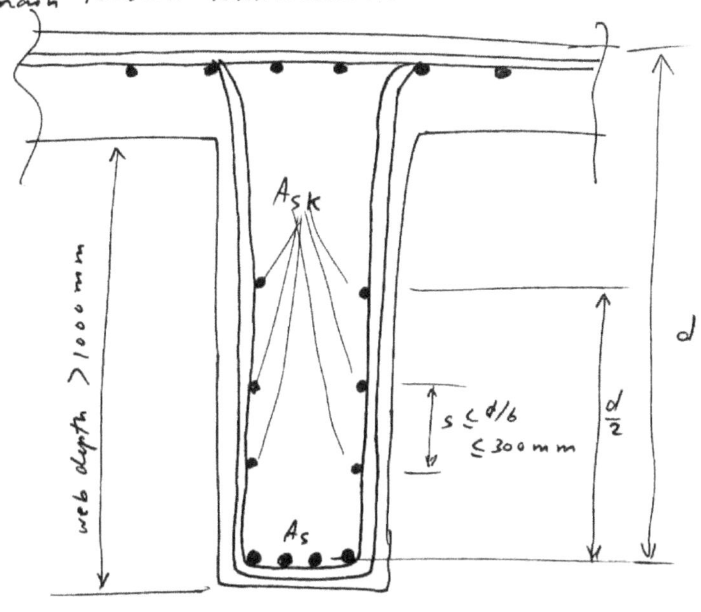

Example 10 :

Design the "skin" reinforcement according to the ACI Code for a rectangular beam 450 mm wide by 1200 mm deep, having 10-⌀28 mm bars as tension reinforcement. Take $f_c' = 28\ MPa$ and $f_y = 415\ MPa$.

Solution :

(a) Establish whether or not skin reinforcement is needed

In this case, the web is the overall depth h.

$$h = 1200\ mm > 1\ m \implies \text{skin reinforcement is needed.}$$

(b) $A_{sk} = 1.0 \, (d - 750)$

$\qquad = 1.0 \, (1140 - 750)$

$\qquad = 390 \text{ mm}^2 \, \left(\dfrac{/ \text{ meter}}{\text{on each side}}\right)$

(c) Spacing:

Maximum spacing is the lesser of $d/6$ and 300 mm.

$\dfrac{d}{6} = \dfrac{1140}{6} = \underline{\underline{190 \text{ mm}}}$

$\therefore$ Use a spacing of 190 mm.

$\Rightarrow$ Use $\phi 10 \text{ mm} @ 190 \text{ mm}$

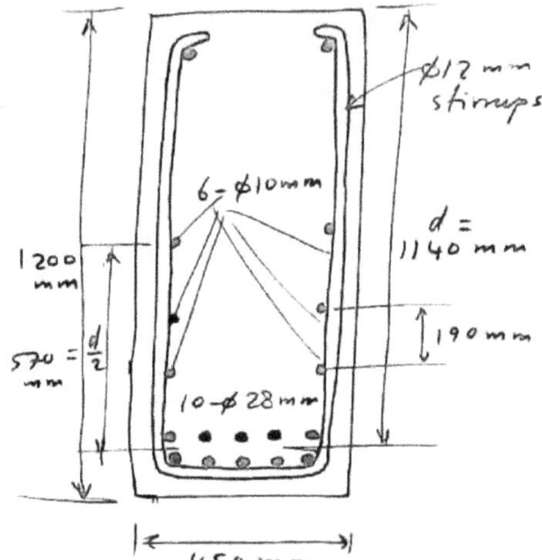

$A_s = 10 \left[\dfrac{\pi (28)^2}{4} \right] = 6{,}158 \text{ mm}^2$

$\Rightarrow A_{sk} \text{ (used)} = 413 \text{ mm}^2 >$ required $A_{sk} = 390 \text{ mm}^2$.

$\Rightarrow$ place within a distance of $\dfrac{d}{2} = 570 \text{ mm}$ nearest the main tension steel.

Total $A_{sk} = 2 \times 413 = 826 \text{ mm}^2$ (on both sides).

Total $A_{sk} = 826 \text{ mm}^2 < \dfrac{1}{2} A_s = \dfrac{1}{2} (6158) = 3079 \text{ mm}^2.$

Chapter 5
Shear Strength and Shear Reinforcement
Part I

5.1 Introduction:

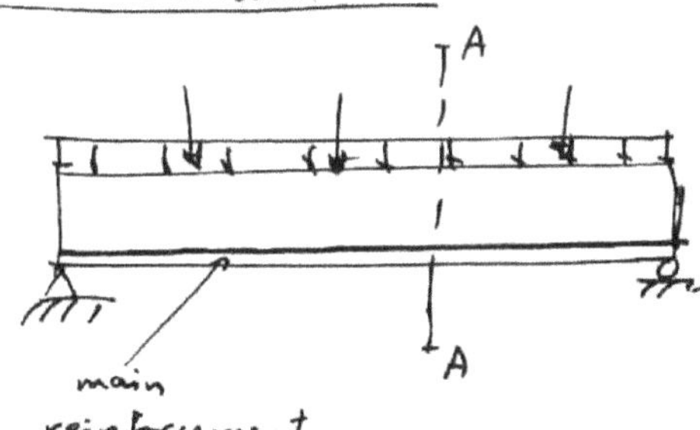
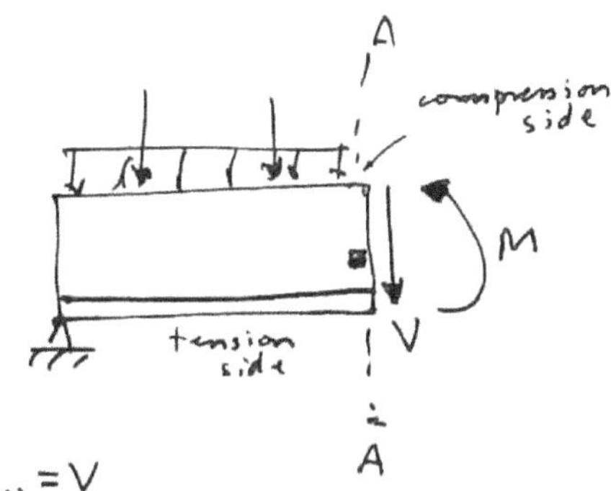

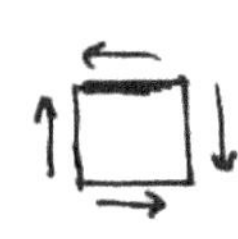

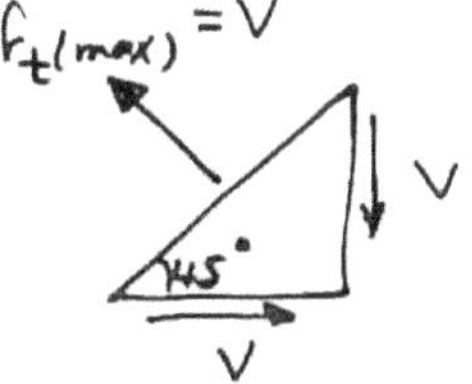

* At the pure shear location, a tensile stress $f_{t(max)}$ is developed equal to the shear stress, and acts on a 45°-plane

* This diagonal tension constitutes the main cause of inclined cracking.

* Shear failures in beams are actually tension failures at the inclined cracks.

5.2 The Shear Stress Formula Based on Linear Stress Distribution:

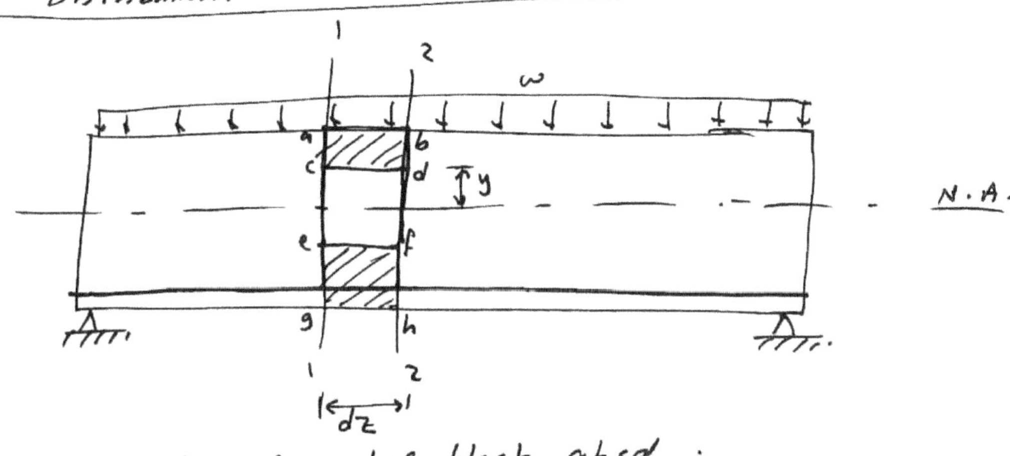

* Consider the elemental block $abcd$:

$\xrightarrow{+} \Sigma F_x = 0:$

$C_1 + v_y\, b\, dz - C_2 = 0$

$v_y\, b\, dz = C_2 - C_1$

where

v_y: horizontal shear stress
  at a distance y
  from the neutral axis.

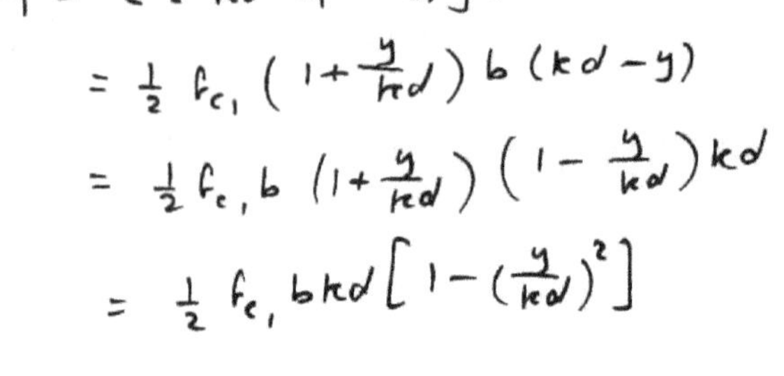

$C_1 = \tfrac{1}{2}\left(f_{c_1} + f_{c_1 y}\right) b\,(kd - y)$

but $f_{c_1 y}$ is obtained from similar triangles:

$$\frac{f_{c_1 y}}{f_{c_1}} = \frac{y}{kd}$$

$$\Rightarrow\quad f_{c_1 y} = \frac{y}{kd}\, f_{c_1}$$

$$\Rightarrow C_1 = \tfrac{1}{2}\left[\frac{y}{kd} f_{c_1} + f_{c_1}\right] b\,(kd - y)$$

$$= \tfrac{1}{2} f_{c_1}\left(1 + \frac{y}{kd}\right) b\,(kd - y)$$

$$= \tfrac{1}{2} f_{c_1}\, b\left(1 + \frac{y}{kd}\right)\left(1 - \frac{y}{kd}\right) kd$$

$$= \tfrac{1}{2} f_{c_1}\, b\,kd\left[1 - \left(\frac{y}{kd}\right)^2\right]$$

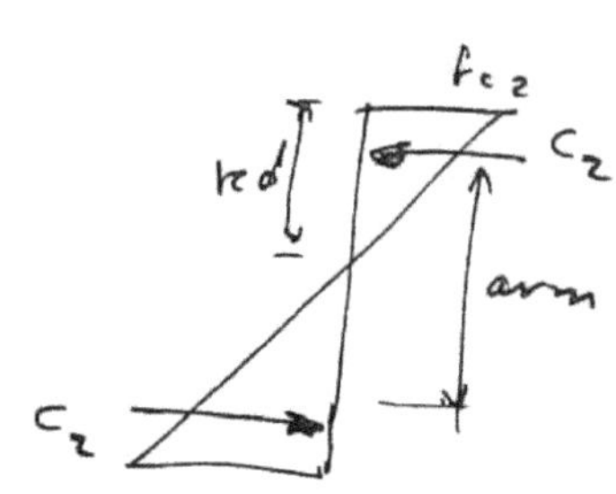

$M_2 = C_2\,(arm)$

$\qquad = \tfrac{1}{2} f_{c_2}\, b\, kd\,(arm)$

$$\Rightarrow\quad \frac{M_2}{arm} = \tfrac{1}{2} f_{c_2}\, b\, kd$$

similarly, $\dfrac{M_1}{arm} = \tfrac{1}{2} f_{c_1}\, b\, kd$

But $\quad C_1 = \left(\tfrac{1}{2} f_{c_1}\, b\, kd\right)\left[1 - \left(\frac{y}{kd}\right)^2\right]$

$$\Rightarrow\quad C_1 = \frac{M_1}{arm}\left[1 - \left(\frac{y}{kd}\right)^2\right]$$

similarly, $\quad C_2 = \dfrac{M_2}{arm}\left[1 - \left(\frac{y}{kd}\right)^2\right]$

Substitute in $\quad v_y\, b\, dz = C_2 - C_1$

$$v_y \, b \, dz = \frac{M_2}{arm}\left[1-\left(\frac{y}{kd}\right)^2\right] - \frac{M_1}{arm}\left[1-\left(\frac{y}{kd}\right)^2\right]$$

$$= \left(\frac{M_2-M_1}{arm}\right)\left[1-\left(\frac{y}{kd}\right)^2\right]$$

$$\Rightarrow v_y = \left(\frac{M_2-M_1}{dz}\right)\frac{1}{b(arm)}\left[1-\frac{y^2}{(kd)^2}\right]$$

but $\quad (M_2-M_1) = V\cdot dz \Rightarrow \dfrac{M_2-M_1}{dz} = V$

$$\Rightarrow \boxed{v_y = \frac{V}{b(arm)}\left[1-\frac{y^2}{(kd)^2}\right]} \quad , \quad 0 \le y \le kd$$

* y is the distance from the neutral axis to any point in the compression area.

* the maximum shear stress v occurs at the neutral axis (i.e. $y=0$).

$$\therefore v = v_{y\,max} = \frac{V}{b(arm)}\left[1-\frac{0}{(kd)^2}\right] = \frac{V}{b(arm)} .$$

* The horizontal shear stress v is also the vertical shear stress, because shear stresses on two perpendicular planes must be equal.

* A state of pure shear exists in regions where the bending stress is low or where flexural cracks exist.

* In this case (pure shear), the maximum principal stress is a tensile stress acting at $45°$, is equal to the shear stress.

* In strength design, the factored shear force V_u is used, and the distribution of flexural stress is no longer linear in the compression zone.

* the tensile stress v is **not** the actual tensile stress.

* In the ACI Code, the denominator in the above equation is replaced by bd .

5.3 The Combined Stress Formula:

* In a homogeneous beam, consider:

f_t : tensile stress

v : shear stress

* Determine the principle tensile stress $f_{t(max.)}$:

* Consider equilibrium of the triangular element:

$\xrightarrow{+} \Sigma F_x = 0$:

$$f_t' \cos\alpha \left(\frac{b\,dz}{\sin\alpha}\right) + v'\sin\alpha \left(\frac{b\,dz}{\sin\alpha}\right)$$

$$- v(b\,dz) - f_t \left(\frac{b\,dz}{\tan\alpha}\right) = 0$$

$\underbrace{\qquad\qquad\qquad\qquad}$ (1)

$+\uparrow \Sigma F_y = 0$:

$$-f_t' \sin\alpha \left(\frac{b\,dz}{\sin\alpha}\right) + v'\cos\alpha \left(\frac{b\,dz}{\sin\alpha}\right) + v\left(\frac{b\,dz}{\tan\alpha}\right) = 0 \qquad\text{---} (2)$$

Solve the above two equations simultaneously for f_t' and v':

$$f_t'\,b\,dz\,\cos\alpha + v'\,b\,dz\,\sin\alpha - v\,b\,dz\,\sin\alpha - f_t\,b\,dz\,\cos\alpha = 0 \quad\text{---}(1)$$

$$-f_t'\,b\,dz\,\sin\alpha + v'\,b\,dz\,\cos\alpha + v\,b\,dz\,\cos\alpha = 0 \qquad\text{---}(2)$$

---

* Eliminate $b\,dz$ from each equation:

$\sin\alpha \Big/ \quad f_t'\cos\alpha + v'\sin\alpha - v\sin\alpha - f_t\cos\alpha = 0 \qquad\text{---}(1)$

$\cos\alpha \Big/ \quad -f_t'\sin\alpha + v'\cos\alpha + v\cos\alpha = 0 \qquad\text{---}(2)$

$$f_t'\sin\alpha\cos\alpha + v'\sin^2\alpha - v\sin^2\alpha - f_t\sin\alpha\cos\alpha = 0 \qquad\text{---}(1)$$

$$-f_t'\sin\alpha\cos\alpha + v'\cos^2\alpha + v\cos^2\alpha = 0 \qquad\text{---}(2)$$

$$v' + v(\cos^2\alpha - \sin^2\alpha) - f_t\sin\alpha\cos\alpha = 0$$

$$v' + v\cos 2\alpha - \tfrac{1}{2}f_t\sin 2\alpha = 0$$

$$\implies \boxed{v' = \tfrac{1}{2}f_t\sin 2\alpha - v\cos 2\alpha}$$

$$\cos\alpha \ \Big/ \ f_t'\cos\alpha + v'\sin\alpha - v\sin\alpha - f_t\cos\alpha = 0 \quad ——— (1)$$

$$-\sin\alpha \ \Big/ \ -f_t'\sin\alpha + v'\cos\alpha + v\cos\alpha = 0 \quad ——— (2)$$

$$f_t'\cos^2\alpha + v'\sin\alpha\cos\alpha - v\sin\alpha\cos\alpha - f_t\cos^2\alpha = 0 \quad ——— (1)$$

$$f_t'\sin^2\alpha - v'\sin\alpha\cos\alpha - v\sin\alpha\cos\alpha = 0 \quad ——— (2)$$

$$f_t' - 2v\sin\alpha\cos\alpha - f_t\cos^2\alpha = 0$$

$$\Longrightarrow \quad f_t' = f_t\cos^2\alpha + v\sin 2\alpha \qquad , \quad \text{but} \quad \cos^2\alpha = \frac{1+\cos 2\alpha}{2}$$

$$\therefore \quad \boxed{f_t' = \tfrac{1}{2}f_t(1+\cos 2\alpha) + v\sin 2\alpha}$$

*** Find** α_{max} (where f_t is max. and v' is zero):

$$v' = 0 \quad \Longrightarrow \quad \tfrac{1}{2}f_t\sin 2\alpha_{max} - v\cos 2\alpha_{max} = 0$$

$$\tfrac{1}{2}f_t\sin 2\alpha_{max} = v\cos 2\alpha_{max}$$

$$\Longrightarrow \quad \tan 2\alpha_{max} = \frac{2v}{f_t} \equiv \frac{v}{(\tfrac{1}{2}f_t)}$$

$$\therefore \quad f_t'{}_{(max)} = \tfrac{1}{2}f_t\left(1 + \frac{\tfrac{1}{2}f_t}{\sqrt{(\tfrac{1}{2}f_t)^2+v^2}}\right) + v\left(\frac{v}{\sqrt{(\tfrac{1}{2}f_t)^2+v^2}}\right)$$

$$= \tfrac{1}{2}f_t\left[\frac{\sqrt{(\tfrac{1}{2}f_t)^2+v^2} + \tfrac{1}{2}f_t}{\sqrt{(\tfrac{1}{2}f_t)^2+v^2}}\right] + \frac{v^2}{\sqrt{(\tfrac{1}{2}f_t)^2+v^2}}$$

$$= \frac{\tfrac{1}{2}f_t\sqrt{(\tfrac{1}{2}f_t)^2+v^2} + \left((\tfrac{1}{2}f_t)^2 + v^2\right)}{\sqrt{(\tfrac{1}{2}f_t)^2+v^2}}$$

$$= \frac{\sqrt{(\tfrac{1}{2}f_t)^2+v^2}\left[\tfrac{1}{2}f_t + \sqrt{(\tfrac{1}{2}f_t)^2+v^2}\right]}{\sqrt{(\tfrac{1}{2}f_t)^2+v^2}}$$

$$\Longrightarrow \quad \boxed{f_{t\,(max.)} = \tfrac{1}{2}f_t + \sqrt{(\tfrac{1}{2}f_t)^2+v^2}}$$

1) When v is small,

$$\tan 2\alpha_{max.} = 0$$

$$2\alpha_{max.} = 0 \implies \alpha_{max.} = 0$$

$$\therefore \quad f_{t(max)} \approx \frac{1}{2}f_t + \sqrt{\left(\frac{1}{2}f_t\right)^2 + 0}$$

$$\approx f_t \quad (\text{at } \alpha = 0).$$

(2) When f_t is small, $\quad \tan 2\alpha_{max} = \infty$

$$\implies 2\alpha_{max} = 90° \implies \alpha_{max} = 45°.$$

$$\text{and} \quad f_{t(max.)} = 0 + \sqrt{0 + v^2} \approx v \quad (\text{at } \alpha = 45°).$$
$$(\underline{\text{inclined cracks}})$$

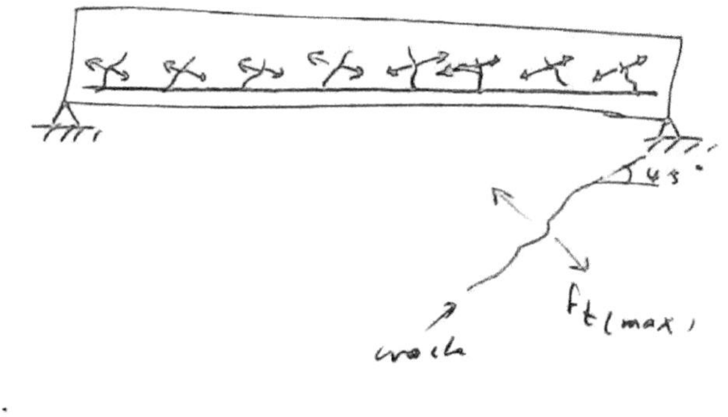

5.4 Behavior of Beams Without Shear Reinforcement:

* To control inclined cracks, <u>transverse reinforcement</u>, also called <u>shear reinforcement</u>, in the form of closed or U-shaped stirrups is used.

* <u>Web-shear Crack</u> : is an inclined crack occurring in a beam that was previously uncracked due to flexure.

* <u>Flexure-shear Crack</u> : is an inclined crack originating at the top of and becoming an extension to a previously existing flexural crack.

* The critical flexural crack is referred to as the <u>initiating crack</u>

Web-shear Crack   Flexure-shear Crack

* Flexure-shear cracks are the usual type found in in reinforced concrete.

* The transfer of shear in reinforced concrete beams occurs by a combination of the following mechanisms:

(1) Shear resistance of the uncracked concrete, V_{cz}.

(2) Aggregate interlock (or interface shear transfer) force, V_a, tangentially along a crack.

(3) Dowel action, V_d, the resistance of the longitudinal reinforcement to a transverse force.

(4) Arch action on relatively deep beams.

(5) Shear reinforcement resistance, V_s, from stirrups (**not** available in beams without shear reinforcement).

* For rectangular beams without shear reinforcement, the proportion of the shear transferred by the various mechanisms is as follows:

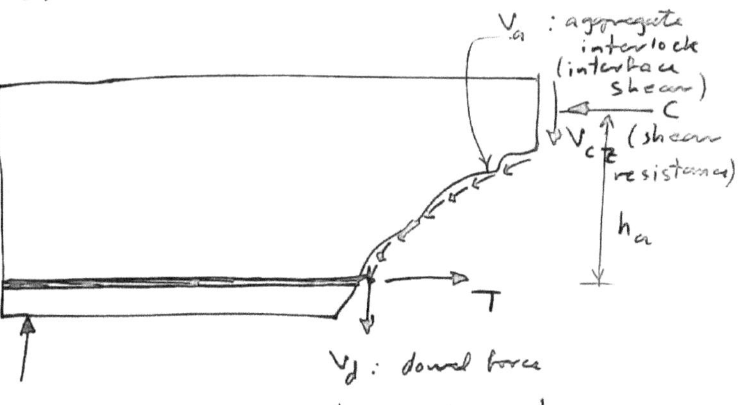

"Redistribution of Shear Resistance After Formation of Inclined Crack"

(1) 15–25% dowel action.

(2) 20–40% uncracked concrete compression zone.

(3) 33–50% aggregate interlock (interface shear transfer).

---

* Inclined cracks begin and grow depending on the relative magnitudes of shear stress v and flexural stress f_t:

$$v = k_1 \frac{V}{bd}$$

$$f_t = k_2 \frac{M}{bd^2}$$

where k_1, k_2 are proportionality constants.

$$\therefore \quad \frac{f_t}{V} = \frac{k_2 \frac{M}{bd^2}}{k_1 \frac{V}{bd}} = \frac{k_2}{k_1} \frac{M}{Vd} = k_3 \frac{M}{Vd}$$

* For the simply-supported beam shown,

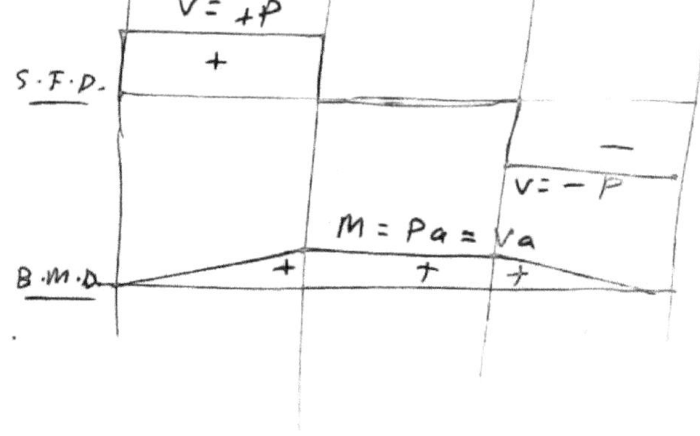

$$\frac{M}{V} = \frac{Va}{V} = a .$$

a : distance over which the shear is constant.

$a \equiv$ "shear span"

* In general,

$$a = \frac{M}{V} \quad (\text{shear span}) .$$

$$\boxed{\frac{f_t}{V} = k_3 \left(\frac{a}{d} \right)}$$

$\dfrac{a}{d} = \dfrac{\text{shear span}}{\text{depth}}$ is a factor in shear strength.

---

* Regarding $\frac{a}{d}$, beams are classified into <u>four</u> general categories of failure:

---

(1) $\frac{a}{d} \leq 1$ $\implies$ <u>deep</u> beams.

(2) $1 < \frac{a}{d} \leq 2\frac{1}{2}$ $\implies$ <u>short</u> beams (shear strength exceeds the inclined cracking capacity).

(3) $2\frac{1}{2} < \frac{a}{d} \leq 6$ $\implies$ <u>usual</u> beams of intermediate length (shear strength equals the inclined cracking capacity).

(4) $\frac{a}{d} > 6$ $\implies$ <u>long</u> beams (flexural strength < shear strength).

Flexural moment strength

shear compression strength

Failure Moment $= V_a$

inclined cracking strength

deep beams

flexural failures

a/d

1   2   3   4   5   6

shear–tension and shear–compression failures

diagonal–tension failures

---

tension tie

compression arch

① <u>Deep Beams</u>: $\frac{a}{d} \le 1$.

* shear stress has the predominant effect.

* After inclined cracking occurs, the simply-supported beam tends to behave like a <u>tied-arch</u>, where the load is carried by direct compression (shaded area) and by the tension in the longitudinal steel.

* Several modes of failure are possible:

(1) <u>Anchorage failure</u> : pullout of the tension reinforcement at the support.

(2) <u>Crushing (Bearing) failure</u> : at the reactions.

(3) <u>Flexural failure</u> : arising from either crushing of concrete near the top of the arch or yielding of tension reinforcement.

(4) <u>Arch-rib failure</u> : due to an eccentricity of the arch thrust (resulting in either a tension crack over the support or a crushing of concrete on the underside of the rib).

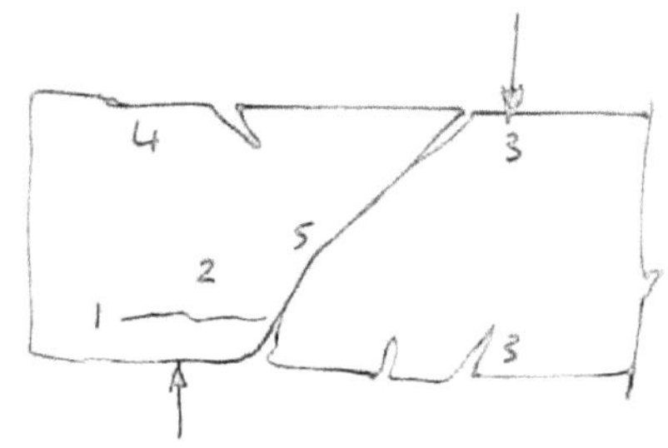

(2) <u>Short Beams</u> : $1 < \frac{a}{d} \leq 2\frac{1}{2}$

* shear strength > inclined cracking strength.
* After the flexure-shear crack develops, the crack extends further into the compression zone as the load increases.
* It also propagates as a secondary crack toward the tension reinforcement and then propagates horizontally along that reinforcement.
* Failure in short beams occurs, either :
    (1) by an anchorage failure at the tension reinforcement (<u>shear-tension</u> failure),
    (2) by a crushing failure in the concrete near the compression face (<u>shear-compression failure</u>).

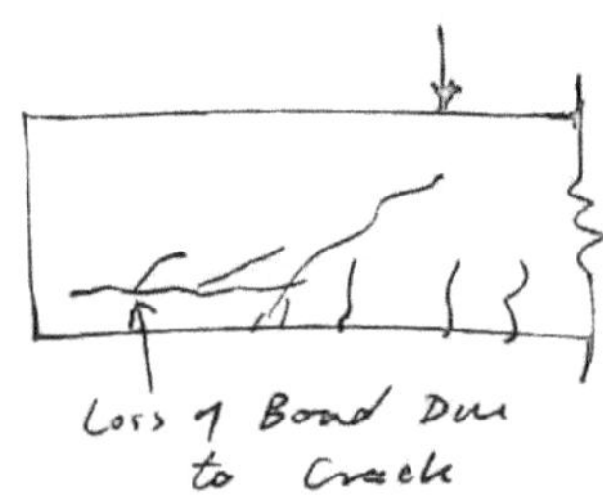

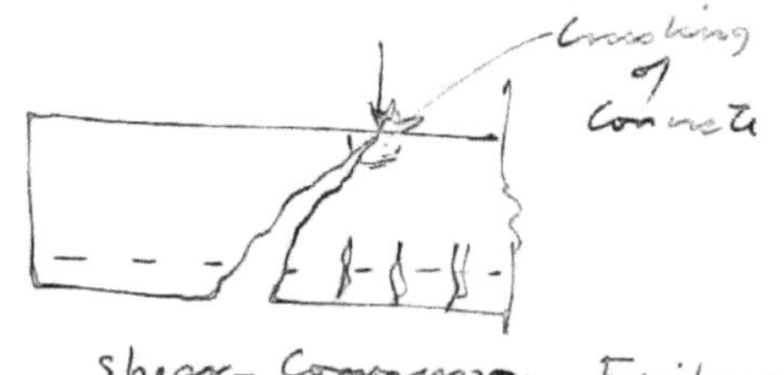

(3) <u>Usual Beams of Intermediate Length:</u> $2\frac{1}{2} < \frac{a}{d} \leq 6$

* Vertical flexural cracks are first to form, followed by the inclined flexure-shear cracks.
* At the beginning, several flexural cracks tend to bend over creating beam segments between cracks (called <u>teeth</u>).
* When the root of the "tooth" is so reduced in size that it becomes unable to carry the moment, it breaks

to form inclined-flexure-shear-cracks.

* At the sudden occurrence of the inclined cracks, the beam is not able to redistribute the load, as in the situation of smaller $\frac{a}{d}$ ratio.

* The formation of the inclined crack represents the shear strength of beams in this category $\Rightarrow$ <u>diagonal-tension failure</u>.

* This is the usual category for beam design.

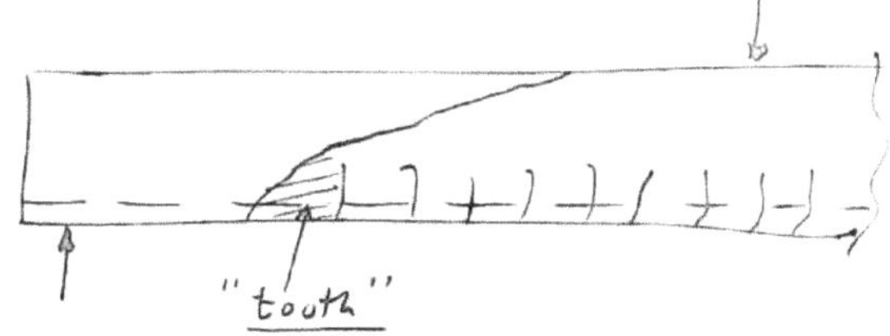

---

④ <u>Long Beams</u>: $\frac{a}{d} > 6$.

* Failure of long beams starts with yielding of the tension reinforcement and ends by crushing of the concrete at the section of maximum bending moment.

* In addition to the nearly vertical flexural cracks at the section of maximum bending moment, prior to failure slightly inclined (from the vertical) cracks may be present between the support and the section of maximum bending moment.

* The strength of the beam is <u>not</u> affected by the size of the shear force.

<u>Summary</u>.

* Shear tends to cause inclined cracks.

* When no such cracks form before the nominal flexural strength is reached, the effect of shear is negligible.

* Beams may fail upon formation of an inclined crack, (diagonal tension failure) when $2\frac{1}{2} \leq \frac{a}{d} < 6$, or there may be considerable reserve strength.

* For the design of all but deep beams, the shear strength is assumed to be reached when the inclined crack forms.

5.5 Shear Strength of Beams <u>without</u> Shear Reinforcement (ACI Code)

* The shear strength of a beam (without shear reinforcement) is the strength at which an inclined crack (flexure–shear crack) forms (ACI Code).

* It is assumed that the strength is reached when the principal tensile stress $f_{t(max)}$ reaches the tensile strength of concrete, which is proportional to $\sqrt{f_c'}$.

$$f_t \propto \frac{E_c}{E_s} f_s \qquad \text{and} \qquad v \propto v_{average}.$$

$$E_c \propto \sqrt{f_c'}$$

M_n : nominal bending moment $\left.\right\}$ at a section when the inclined crack forms.
V_n : nominal shear force

$$v = k_1 \frac{V_n}{bd} \qquad , \qquad f_s \propto \frac{M_n}{A_s d}$$

$$\therefore \ f_t \propto \frac{E_c}{E_s} f_s \propto \frac{E_c}{E_s} \cdot \frac{M_n}{A_s d} \propto \frac{\sqrt{f_c'}}{E_s} \cdot \frac{M_n}{\rho bd \cdot d} = \left(\frac{M_n}{bd^2}\right)\left(\frac{\sqrt{f_c'}}{\rho E_s}\right)$$

$\therefore$ We can write $\quad f_t = \left(\dfrac{k_4}{E_s}\left(\dfrac{\sqrt{f_c'}}{\rho}\right)\right)\dfrac{M_n}{bd^2} \quad \xleftarrow{k_2}$

$$f_{t(max)} = k_5 \sqrt{f_c'}$$

But $\qquad f_{t(max)} = \tfrac{1}{2} f_t + \sqrt{\left(\tfrac{1}{2} f_t\right)^2 + v^2}$

$$k_5\sqrt{f_c'} = \frac{1}{2}\frac{k_4}{E_s}\left(\frac{\sqrt{f_c'}}{\rho}\right)\frac{M_n}{bd^2} + \sqrt{\left[\frac{1}{2}\frac{k_4}{E_s}\left(\frac{\sqrt{f_c'}}{\rho}\right)\frac{M_n}{bd^2}\right]^2 + k_1^2 \frac{V_n^2}{b^2 d^2}}$$

$$k_5\sqrt{f_c'} = \frac{1}{2}\frac{k_4}{E_s}\left(\frac{\sqrt{f_c'}}{\rho}\right)\frac{M_n}{bd^2} + \frac{V_n}{bd}\sqrt{\left[\frac{1}{2}\frac{k_4}{E_s}\left(\frac{\sqrt{f_c'}}{\rho}\right)\frac{M_n}{bd^2}\right]^2 \cdot \frac{b^2 d^2}{V_n^2} + k_1^2}$$

$$k_5\sqrt{f_c'} = \frac{1}{2}\frac{k_4}{E_s}\left(\frac{\sqrt{f_c'}}{\rho}\right)\frac{M_n}{bd^2} + \frac{V_n}{bd}\sqrt{\left[\frac{1}{2}\frac{k_4}{E_s}\left(\frac{\sqrt{f_c'}}{\rho}\right)\left(\frac{M_n}{V_n d}\right)\right]^2 + k_1^2}$$

Multiply the equation by $\dfrac{b d}{V_n}$:

$$k_s \sqrt{f_c'}\,\frac{bd}{V_n} = \frac{1}{2}\frac{k_y}{E_s}\left(\frac{\sqrt{f_c'}}{\rho}\right)\left(\frac{M_n}{V_n d}\right) + \sqrt{\left[\frac{1}{2}\frac{k_y}{E_s}\left(\frac{\sqrt{f_c'}}{\rho}\right)\left(\frac{M_n}{V_n d}\right)\right]^2 + k_1^{\,2}}$$

$$\Longrightarrow \qquad \boxed{\;\frac{V_n}{bd\sqrt{f_c'}} = \frac{k_s}{\dfrac{1}{2}\dfrac{k_y}{E_s}\left(\dfrac{\sqrt{f_c'}}{\rho}\right)\left(\dfrac{M_n}{V_n d}\right) + \sqrt{\left[\dfrac{1}{2}\dfrac{k_y}{E_s}\left(\dfrac{\sqrt{f_c'}}{\rho}\right)\left(\dfrac{M_n}{V_n d}\right)\right]^2 + k_1^{\,2}}}\;}$$

* In the above equation, the variables are

$$\left(\frac{V_n}{bd\sqrt{f_c'}}\right) \quad \text{and} \quad \left(\frac{M_n\sqrt{f_c'}}{E_s\,\rho\,V_n d}\right) \Longrightarrow \text{non-dimensional quantities.}$$

* V_n : nominal shear strength causing the critical inclined crack.

* M_n : corresponding moment at the top of the initiating crack.

* ACI—ASCE committee 326 conducted 440 tests in 1962 to get the following relation:

$$\boxed{\;\frac{V_n}{bd\sqrt{f_c'}} = 1.9 + 2500\,\frac{\rho\,V_n d}{M_n\sqrt{f_c'}} \le 3.5 \quad (f_c' \text{ in psi})\;}$$

* The above equation is acceptable for predicting the flexure-shear cracking load, particularly for $\dfrac{M_n}{V_n d}$ ratios of about $2\frac{1}{2} \to 6$ where $\dfrac{M_n}{V_n d} = \left(\dfrac{M_n}{V_n}\right)/d = a/d$, with considerable conservatism for lower $M_n/(V_n d)$ values.

* Favorable factors for the shear strength of beams without shear reinforcement are:
    (1) a high percentage ρ of longitudinal reinforcement.
    (2) a high ratio of $V_n d/M_n$, i.e. a low a/d ratio.

---

* The ACI Code accepted the above relation since 1963 as the relation defining the shear (inclined cracking) strength of beams without shear reinforcement.

* V_c : nominal shear strength of a beam without shear reinforcement (i.e. $V_n = V_c$) :

$$V_c = \left(1.9\sqrt{f_c'} + 2500\, \frac{\rho_w V_u d}{M_u}\right) b_w d \;\le\; 3.5\sqrt{f_c'}\, b_w d$$

where b_w is the web width. $\qquad\qquad (f_c' \text{ in psi})$

* The <u>web</u> is the width dimension of a beam at its neutral axis.

* In SI Units, we have the formula :

$$V_c = \frac{1}{6}\left(\sqrt{f_c'} + 100\, \frac{\rho_w V_u d}{M_u}\right) b_w d \;\le\; 0.3\sqrt{f_c'}\, b_w d$$

$$(f_c' \text{ in MPa})$$

* The ACI-Code Formula (Section 11.3.2.1) is :

$$V_c = \frac{1}{7}\left(\sqrt{f_c'} + 120\, \frac{\rho_w V_u d}{M_u}\right) b_w d \;\le\; 0.3\sqrt{f_c'}\, b_w d$$

$$(f_c' \text{ in MPa}).$$

<u>Note</u> : Using the factored quantities V_u and M_u instead of $V_n = V_u/\phi$ and $M_n = M_u/\phi$ makes little difference because the ratio $V_u/M_u \approx V_n/M_n$.

<u>Note</u> : $\rho_w = \dfrac{A_s}{b_w d}$, where $b_w =$ web width (<u>not</u> the flange width)

<u>Note</u> : $\dfrac{V_u d}{M_u} \le 1.0$ except when axial compression is present (to be discussed later).

<u>Continuous Beams</u> :

* compression-strut action.

* ACI Committe recommends using $V_c = \frac{1}{6}\sqrt{f_c'}\, b_w d$ (MPa)

$$\boxed{V_c = 2\sqrt{f_c'}\, b_w d}$$

instead of the formula above.

Lightweight Concrete:

* the formula of V_c is slightly modified for lightweight concrete.

* For lightweight concrete, use f_{ct} (tensile strength based on the split cylinder).

* For normal-weight concrete, $f_{ct} = 6.7 \sqrt{f_c'}$ (psi).

$$V_c = \left[1.9 \left(\frac{f_{ct}}{6.7} \right) + 2500 \, \rho_w \frac{V_u d}{M_u} \right] b_w d \leq 3.5 \left(\frac{f_{ct}}{6.7} \right) b_w d$$

and $\quad \dfrac{f_{ct}}{6.7} \leq \sqrt{f_c'}$.

In **SI Units**, use $1.8 f_{ct}$ for $\sqrt{f_c'}$

(ACI Code/ Section 11.2.1.1)

$$V_c = \frac{1}{7} \left(1.8 f_{ct} + 120 \, \rho_w \frac{V_u d}{M_u} \right) b_w d \leq 0.3 \left(1.8 f_{ct} \right) b_w d$$

and $\quad 1.8 f_{ct} \leq \sqrt{f_c'}$. $\qquad$ (in MPa).

Note: $\quad 1.8 f_{ct} = \sqrt{f_c'} \implies f_{ct} = 0.56 \sqrt{f_c'}$, (MPa).

(see chapter 1).

---

* Inclined cracking strength for lightweight concrete varies from about $60\% \to 100\%$ of the values for normal-weight concrete of the same nominal compressive strength.

* Using 0.75 to 0.85 of the tensile strength related term $\sqrt{f_c'}$ for normal-weight concrete in shear-strength equations is a reasonable and generally conservative approach.

* ACI Code/Section 11.2.1.2 permits:

   (1) multiplying all values of $\sqrt{f_c'}$ appearing in shear (or torsion) equations by 0.75 for "all-lightweight" concrete.

   (2) multiplying all values of $\sqrt{f_c'}$ appearing in shear (or torsion) equations by 0.85 for "sand-lightweight" concrete [when the split-cylinder test is not specified].

For "all-lightweight" concrete:

$$V_c = \frac{1}{7}\left[0.75\sqrt{f_c'} + 120\,\rho_w\frac{V_u d}{M_u}\right]b_w d \leq 0.75(0.3)\sqrt{f_c'}\,b_w d$$

For "sand-lightweight" concrete:

$$V_c = \frac{1}{7}\left[0.85\sqrt{f_c'} + 120\,\rho_w\frac{V_u d}{M_u}\right]b_w d \leq 0.85(0.3)\sqrt{f_c'}\,b_w d$$

* Linear interpolation is permitted when partial sand replacement is used.

High Strength Concrete:

* high strength concrete: $f_c' \geq 8000$ psi $(55\ MPa)$.

* shear strength does not increase in proportion to $\sqrt{f_c'}$.

* ACI Code / Section 11.1.2 : $\underset{=}{\underline{8.3\ MPa}}$ (max. $\sqrt{f_c'}$).
  "Limit $\sqrt{f_c'}$ to 100 psi $\left(\frac{25}{3}\,MPa\right)$ in calculations
  of shear strength and development length."

* One exception to this rule (ACI Code / Section 11.1.2.1) is that values of $\sqrt{f_c'}$ greater than $8.3\ MPa$ may be used in computing V_c when A_v is at least $f_c'/35$ times the amount required in ACI Code / Sections 11.5.5.3, 11.5.5.4, or 11.5.5.5 , but no more than three times these amounts.
  i.e. when $\left(\dfrac{b_w s}{3 f_y}\right) \leq A_v \leq \dfrac{f_c'}{35}\left(\dfrac{b_w s}{3 f_y}\right)$.

  $(SI\ Units)$
  b_w, s in $\underline{\underline{mm}}$.

  A_v : Area of shear reinforcement.

5.6 Function of Shear Reinforcement:

* Types of shear reinforcement:
  (1) vertical stirrups (stirrups perpendicular to the longitudinal reinforcement).
  (2) welded wire fabric (wires located perpendicular to the axis of the member).
  (3) inclined stirrups (plane of stirrup makes an angle of $45°$ or more with the longitudinal reinforcement).
  (4) longitudinal tension bars bent toward the compression zone of the member, so that the axis of the bent portion makes an angle of $30°$ or more with the axis of the longitudinal portion.
  (5) combinations of (1) or (3) with (4).

* Spirals are also permitted.

---

* Vertical stirrups are most commonly used.

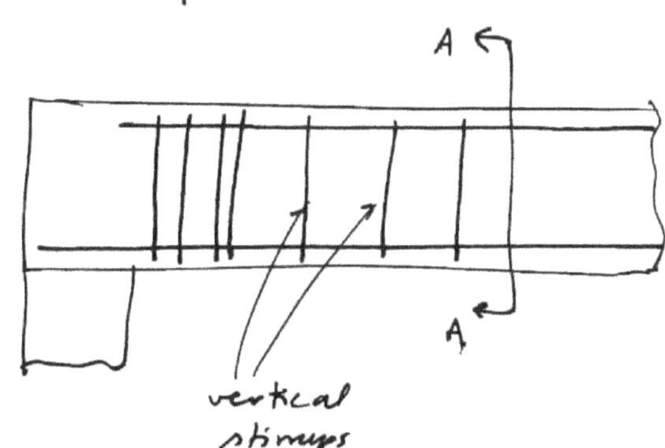

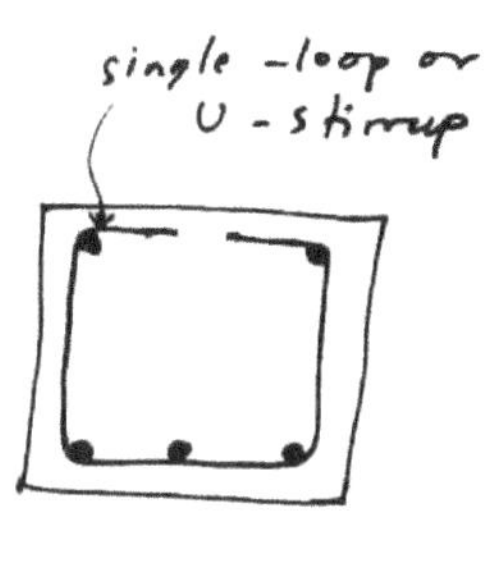

---

* The most generally accepted model for the behavior of reinforced concrete beams containing shear reinforcement is the truss model (Ritter and Mörsch, 1987).
  (presented by Schlaich, Schäfer and Jennewein, 1987).

---

* While shear reinforcement provides _shear strength_, its contribution to the strength occurs only after inclined cracks form.
* Prior to the formation of inclined cracks, the concrete performs the task of carrying the shear.

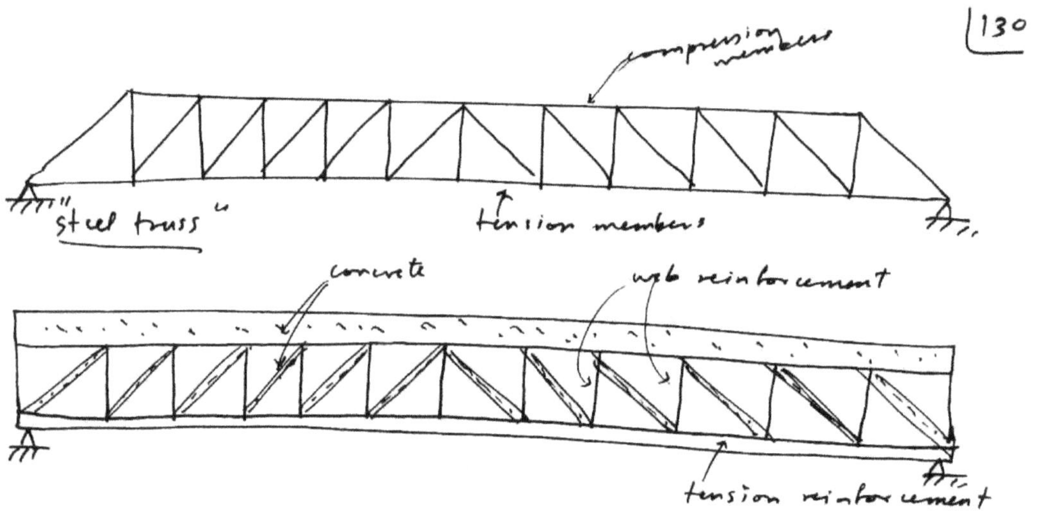

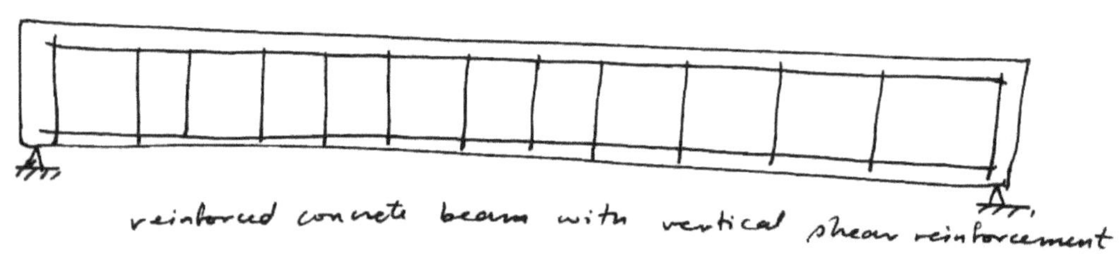

reinforced concrete beam with vertical shear reinforcement

---

* Shear reinforcement is _necessary_ in order to allow a redistribution of internal forces across any inclined crack that may form, and thus prevent a sudden failure upon formation of the crack.

* Shear reinforcement has _several functions_:

  (1) carry part of the shear, V_s.

  (2) restrict the growth of the inclined crack, and thus help maintain aggregate interlock (_interface shear transfer_) strength, V_a

  (3) tie the longitudinal bars in place and thereby increase their strength V_d to resist transverse forces (_dowel action_).

---

* If the amount of shear reinforcement is _too little_, it will yield immediately at the formation of an inclined crack, and the beam then fails.

* If the amount of shear reinforcement is _too much_, there will be a shear-compression failure without the yielding of the shear reinforcement.

* The optimum amount should be such that both the shear reinforcement and the compression zone of the beam each continue to carry increasing shear after the formation of the inclined crack, until the shear reinforcement yields, resulting in a ductile failure.

5.7 Truss Model for Reinforced Concrete Beam:

s : spacing of vertical stirrups.

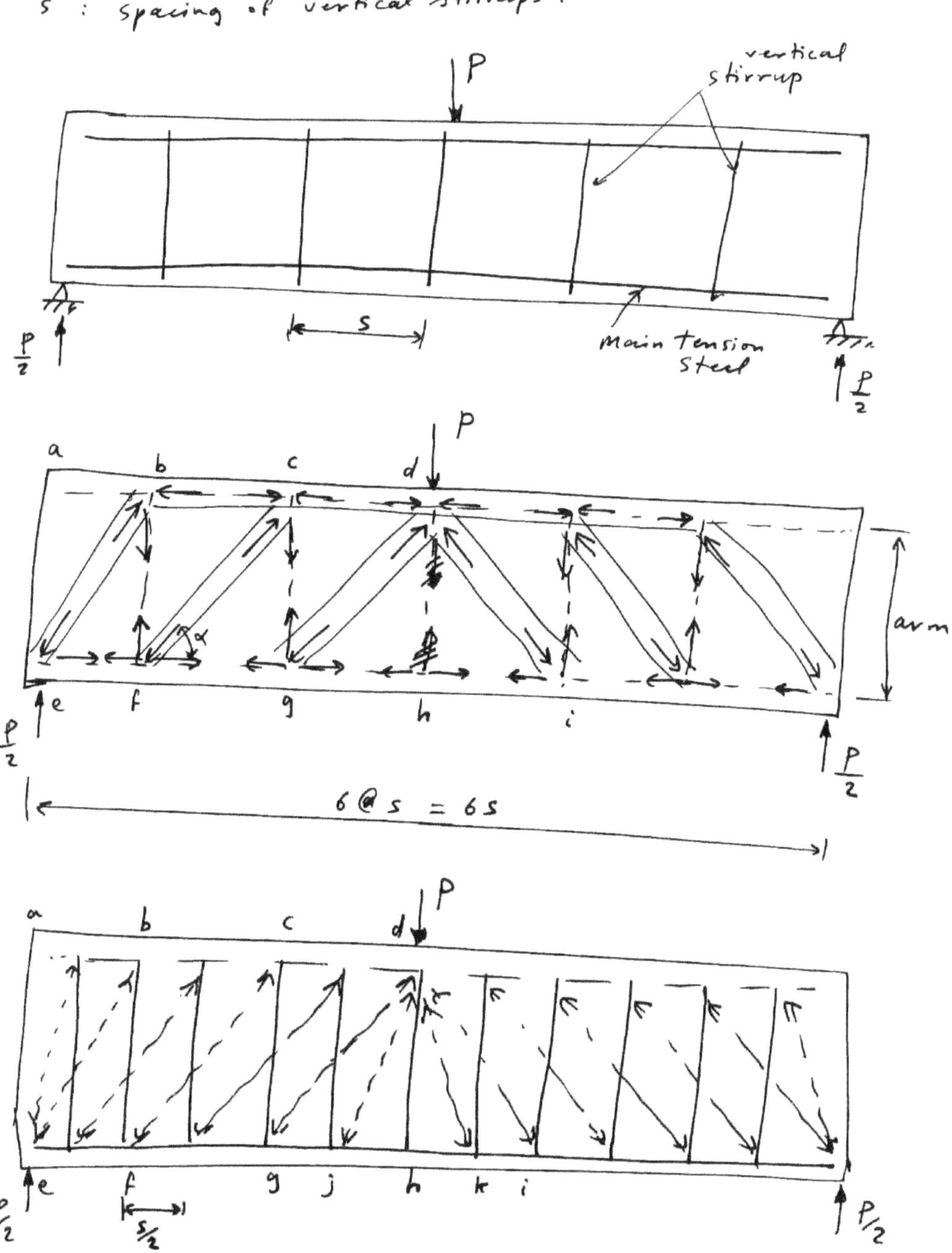

* truss model $\Rightarrow$ compression field theory.
  $\Rightarrow$ strut-and-tie model:

* <u>strut</u>: compression member.
* <u>tie</u>: tension member.

* For stirrups spaced at S, the bending moment should correspond to the "jumped" diagram shown in the figure.

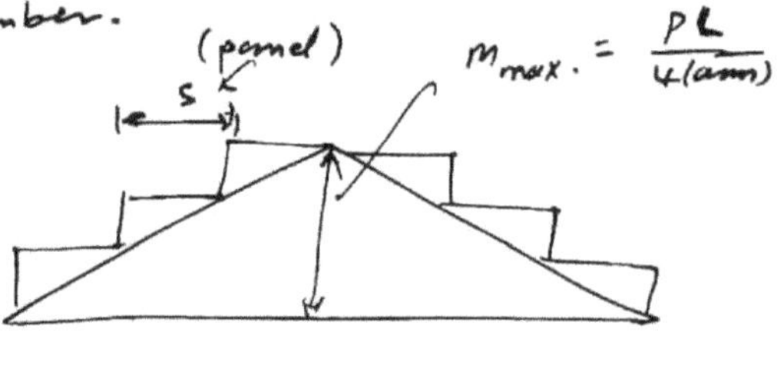

* the tensile force in the truss will be constant over the panel distance; thus, it will change only at a stirrup.

$$\underline{B.M.D.}$$
$$M_{max.} = \frac{PL}{4}.$$
$$Force = \frac{PL}{4(arm)}$$

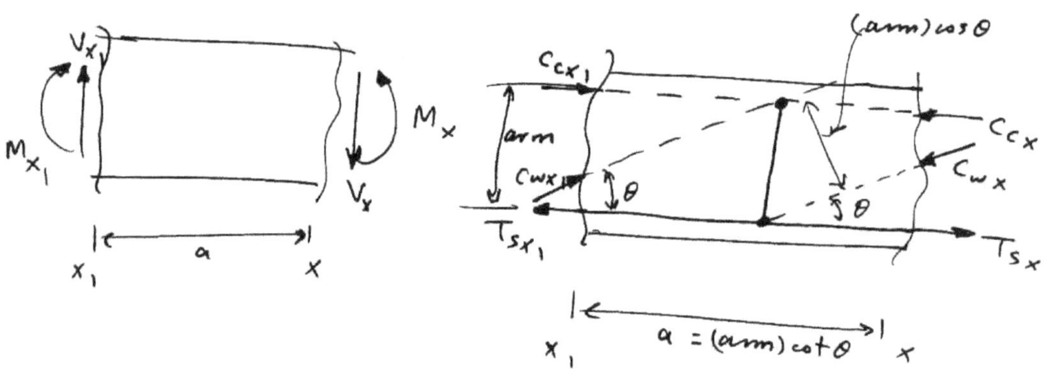

C_{cx} : concrete compression chord force.

C_{wx} : diagonal compression-field "strut" force.

T_{sx} : tensile force carried by the main tension steel.

$$C_{cx} = \frac{M_x}{arm} - \frac{1}{2} V_x \cot\theta$$

$$T_{sx} = \frac{M_x}{arm} + \boxed{\frac{1}{2} V_x \cot\theta} \Longrightarrow$$

additional amount larger than that obtained from the bending moment diagram.

$$C_{wx} = \frac{V_x}{\sin\theta}$$

* this effect has been accounted for in the ACI Code design by extending the bars a <u>distance</u> d beyond the point

where the calculated requirement is made.

* It is clear that shear reinforcement behavior and the forces in longitudinal bars cannot be entirely separated. (See Chapter 6 on development length).

---

* Distribution of Internal Shears in Beams with Shear Reinforcement:

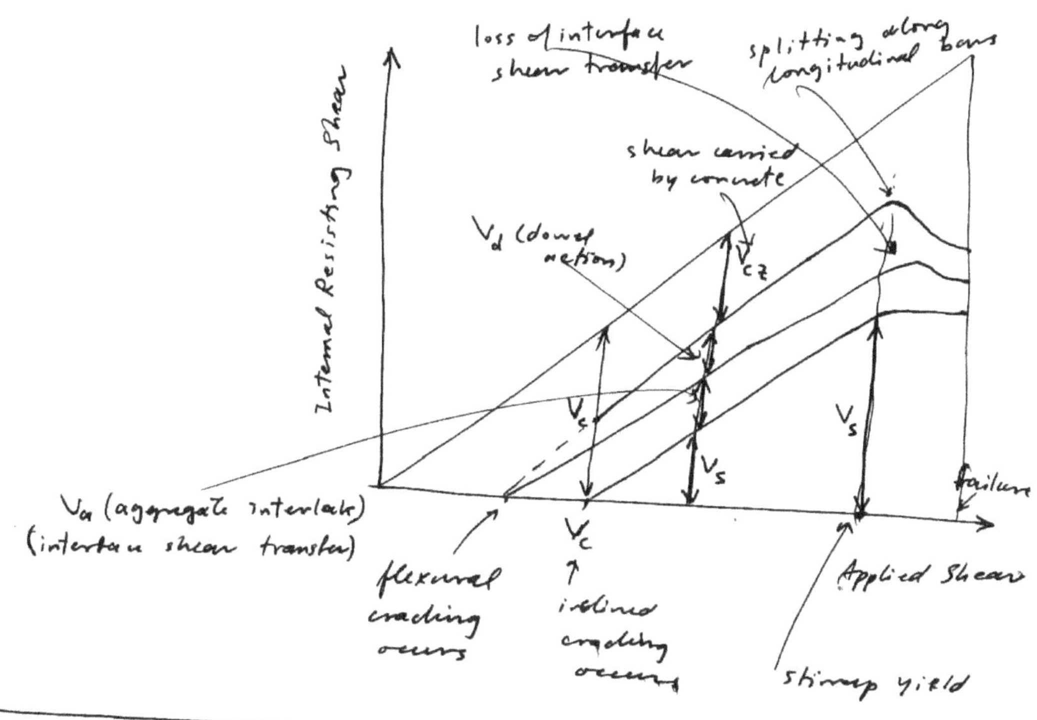

5.8 Shear Strength of Beams with Shear Reinforcement — ACI Code:

V_n: total nominal shear $(V_u \le \phi V_n)$.

$$V_n = V_c + V_s$$

V_c: shear strength due to concrete

V_s : shear strength due to shear reinforcement.

* Determine an expression for V_s from the truss analogy.

* Assume an inclined crack at 45° extends across the beam intersecting an average of N shear reinforcing bars

A_v: area of shear reinforcement within a distance s.

f_y: tensile yield stress for the shear reinforcement

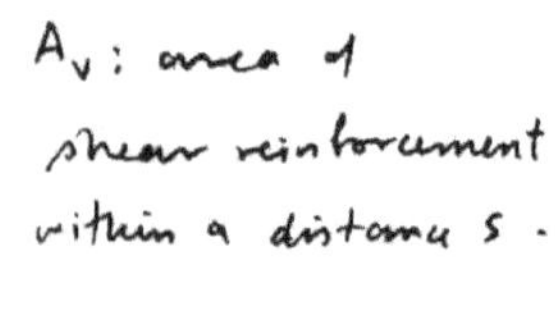

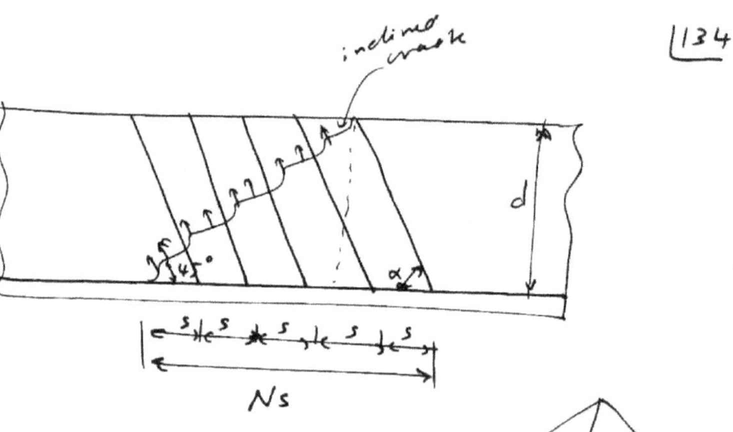

$$V_s = N A_v f_y \sin\alpha$$

But $\quad \tan 45° = \dfrac{d}{x_1} \implies x_1 = d\cot 45°$

$\qquad \tan\alpha = \dfrac{d}{x_2} \implies x_2 = d\cot\alpha$

$\implies Ns = x_1 + x_2 = d(\cot 45° + \cot\alpha) = d(1 + \cot\alpha)$

$\implies V_s = \dfrac{d(1+\cot\alpha)}{s} A_v f_y \sin\alpha = \dfrac{A_v f_y (\sin\alpha + \cos\alpha) d}{s}$

* For vertical stirrups, $\alpha = 90°$

$\implies V_s = \dfrac{A_v f_y (1+0) d}{s} = \dfrac{A_v f_y d}{s}$

$$\implies \boxed{V_s = \dfrac{A_v f_y d}{s}} \quad \bigstar$$

5.9 Lower and Upper Limits of Shear Reinforcement:

* Optimal amount of shear reinforcement should be used in order to ensure yielding of steel when the failure strength in shear is reached.

* ACI Code (Formula 11-14) / Section 11.5.5.3 :

$$\underline{\underline{\text{min.}} \; A_v = \dfrac{50\, b_w s}{f_y}} \quad psi \qquad \left(b_w, s \;\text{in in.} \atop f_y \;\text{in} \; \boxed{psi} \right)$$

$$\underline{SI\ Units\,;} \quad \min A_v = \dfrac{b_w s}{3 f_y} \; SI \qquad \left(b_w, s \;\text{in mm} \atop f_y \;\text{in MPa} \right).$$

where A_v in mm^2 and b_w is the beam web width. $\boxed{135}$

$$V_s = A_v \frac{f_y d}{s} \geq \frac{f_y d}{s}\left(\frac{b_w s}{3 f_y}\right) = \frac{1}{3} b_w d$$

$$\Rightarrow \boxed{V_s \geq \frac{1}{3} b_w d} \qquad (V_s \text{ in } MN).$$

shear stress $v_s = \dfrac{V_s}{b_w d} \geq \dfrac{1}{3} \qquad (MPa)$.

* Maximum Value of V_s :

$$V_s \leq \frac{1}{2}\sqrt{f_c'}\, b_w d \quad \text{to} \quad \frac{2}{3}\sqrt{f_c'}\, b_w d$$

* ACI Code / Section 11.5.6.8 : $V_{s_{max.}} = \frac{2}{3}\sqrt{f_c'}\, b_w d$.

5.10 Critical Section for Nominal Shear Strength Calculation:

* In experimental work, the critical section for computing the nominal shear strength was the location of the first inclined crack.

* Development of the empirical equation of V_c was based on:

(1) For $\frac{a}{d} > 2$, the critical inclined crack is expected at d from the section of maximum moment.

(2) For $\frac{a}{d} < 2$, an inclined crack is expected at the center of the shear span.

* For gradually varying shear force (e.g. uniform loads), ACI Code / Section 11.1.3 permits taking the critical section at a distance d from the face of support.

* Shear reinforcement must be provided between the face of support and the distance d therefrom, using the same requirements as at the critical section.

* The critical section must be taken at the face of support when one of the following occurs:

(1) Factored shear V_u does gradually decrease from the face of support but the support is itself a beam or girder and therefore does **not** introduce compression into the end region of the member.

(2) When a concentrated load occurs between the face of support and the distance d therefrom.

(3) When any loading may cause a potential inclined crack **at** the face of support or **extend into** instead of away from the support.

inclined crack extending
into the support.

5·11 ACI Code Provisions for Shear Strength of Beams:

* ACI Strength Design Method: $\boxed{V_u \leq \phi V_n}$

  V_u : factored shear force

  ϕV_n : design shear strength.

  $\boxed{\phi = 0.85 \quad \text{for shear}}$,

$$\boxed{V_n = V_c + V_s}$$

Strength V_c due to Concrete:

① the **simplified** method: $\qquad$ IF $s_w > 0.012$

$$V_c = \frac{1}{6} \sqrt{f_c'} \, b_w d$$

, f_c' in MPa
b_w, d in mm
V_c in MN .

V_c: nominal shear strength of a beam without shear reinforcement.

* The above value appears to be conservative; however, recent studies have shown otherwise when ρ_w is below 0.012, where $\rho_w = \dfrac{A_s}{b_w d}$.

* For values of $\rho_w < 0.012$, the following is suggested:

$$V_c = (0.07 + 8.3\,\rho_w)\sqrt{f_c'}\; b_w d$$

② the __more detailed method__:

$$V_c = \frac{1}{6}\left(\sqrt{f_c'} + 100\,\frac{\rho_w V_u d}{M_u}\right) b_w d \;\leq\; 0.3\sqrt{f_c'}\; b_w d$$

* The value $\dfrac{V_u d}{M_u} \leq 1.0$, (ACI Code / Section 11.3.2.1)

* M_u is the factored moment occurring simultaneously with the V_u for which the shear strength is being provided.

Strength V_s due to Shear Reinforcement:

* ACI Code / Section 11.5.6 :
$$V_s = \frac{A_v f_y d}{s}(\sin\alpha + \cos\alpha)$$

* For vertical stirrups $(\alpha = 90°)$, $\quad V_s = \dfrac{A_v f_y d}{s}$

Design Categories and Requirements:

* It is better to plot the V_u diagram rather than the "required V_n" diagram.

* The design for shear may be separated into the following categories:

① $V_u \leq 0.5\,\phi V_c \implies$ no shear reinforcement is required.
    (ACI Code / Section 11.5.5.1).

② $0.5\,\phi V_c < V_u \leq \phi V_c$

$\implies$ Minimum shear reinforcement is required except for thin slablike flexural members.

* The thin slablike members include :

    (a) slabs
    (b) footings
    (c) floor joist construction
    (d) beams where total depth $\leq$ 10 in.(250 mm), i.e.
        $2\frac{1}{2}$ times the flange thickness for T-sections,
        or one-half of the web width.

* ACI Code / Section 11.5.5.3 and 11.5.4.1 :

$$\text{required } \phi V_s = \min \phi V_s = \phi \left(\tfrac{1}{3}\right) b_w d$$

and maximum spacing $s \leq \dfrac{d}{2} \leq 24$ in (600 mm).

---

③ $\phi V_c < V_u \leq \left[\phi V_c + \min \phi V_s \right]$

* In this case, for all flexural members, including those
exempted from shear reinforcement in Category 2, shear
reinforcement must be provided satisfying :

$$\text{required } \phi V_s = \min \phi V_s = \tfrac{1}{3} \phi\, b_w d$$

    maximum spacing $s \leq \dfrac{d}{2} \leq 600$ mm.

---

④ $\left[\phi V_c + \min \phi V_s \right] < V_u \leq \left[\phi V_c + \phi\left(\tfrac{1}{3}\sqrt{f_c'}\right) b_w d \right]$

* For this category, the computed shear reinforcement
will exceed the min. ϕV_s requirement, and the shear
reinforcement must satisfy ACI Formula (11-2),
ACI Code / Sections 11.5.6 , 11.5.4.1 and 11.5.4.3 :

$$\text{required } \phi V_s = V_u - \phi V_c$$

$$\text{provided } \phi V_s = \frac{\phi A_v f_y d}{s} \qquad (\text{for } \alpha = 90°).$$

    maximum $s = \dfrac{d}{2} \leq 600$ mm.

<u>Note</u> : In terms of nominal stress, $v_s = \dfrac{V_s}{b_w d} = \tfrac{1}{3}\sqrt{f_c'}$ MPa
is the maximum v_s for which the $\dfrac{d}{2}$ maximum
spacing limit applies.

⑤ $\left[\phi V_c + \phi\left(\frac{1}{3}\sqrt{f_c'}\right)b_w d\right] < V_u \leq \left[\phi V_c + \phi\left(\frac{2}{3}\sqrt{f_c'}\right)b_w d\right]$

* The difference between categories 4 and 5 is that for all regions of a beam where the nominal stress v_s to be taken by shear reinforcement is between $\frac{1}{3}\sqrt{f_c'}$ and $\frac{2}{3}\sqrt{f_c'}$, the maximum shear reinforcement spacing s may not exceed $d/4$.

* The shear reinforcement in this category must satisfy ACI Formula (11-2), ACI Code/Sections 11.5.6, 11.5.4.3, and 11.5.6.8 :

$$\text{required } \phi V_s = V_u - \phi V_c$$

$$\text{provided } \phi V_s = \frac{\phi A_v f_y d}{s} \qquad (\text{for } \alpha = 90°)$$

$$\text{maximum spacing } \quad s \leq \frac{d}{4} \leq 300 \text{ mm.}$$

* The factored shear V_u may not exceed the upper limit $\left[\phi V_c + \phi\left(\frac{2}{3}\sqrt{f_c'}\right)b_w d\right]$ according to ACI Code/Section 11.5.6.8.

* The maximum factored shear that must be provided for any beam is that occurring at the <u>critical section</u>.

* The V_u requirement between the face of the support and the critical section is to be taken as constant, equal to the value at the critical section.

5.12 Working Stress Method - ACI Code, Appendix A :

* The nominal shear stress in the "alternate design method" is based on the service load shears without applying the overload factors U or the strength reduction factors ϕ ; thus :

$$v = \frac{V_D + V_L}{b_w d} \qquad (\text{shear stress}).$$

* Allowable shear stresses are taken as:

$\begin{cases} 55\% \text{ of stresses in strength design, for beams, joists,} \\ \qquad \text{walls, and one-way slabs.} \\ 50\% \text{ of stresses in strength design, for two-way slabs} \\ \qquad \text{and footings.} \end{cases}$

* For beams,

$$\text{allowable } v_c = \frac{V_c}{b_w d} = \frac{55}{100} \cdot \frac{1}{6}\sqrt{f_c'} = 0.092\sqrt{f_c'}.$$

$$\approx \frac{\sqrt{f_c'}}{11}.$$

(ACI Code / Section A3.1).

$$\cong \text{ allowable } v_c = \frac{\sqrt{f_c'}}{12} + 9\rho_w \frac{Vd}{M} \leq \frac{\sqrt{f_c'}}{7}.$$

(ACI Code / Section A7.4.4).

* the working stress method is <u>not</u> used in the design of shear reinforcement.

5.13 Shear Strength of Beams — Design Examples:

* shear strength design $\Longrightarrow$ design of vertical stirrups.

Example 1 :

For the given beam shown in the figure, first determine the maximum uniform dead and live loads under service condition permitted by the ACI strength method. Then using those maximum service loads, design the shear reinforcement using vertical stirrups and the strength method with the simplified procedure using constant V_c. Assume that service live load to dead load ratio is 1.5, $f_c' = 28$ MPa, and $f_y = 415$ MPa.

Solution :

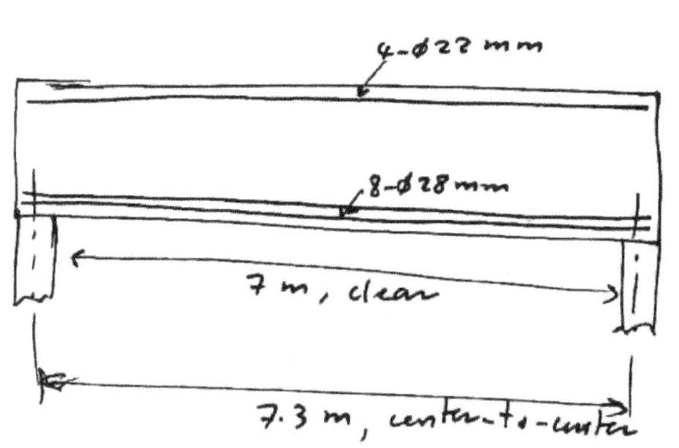

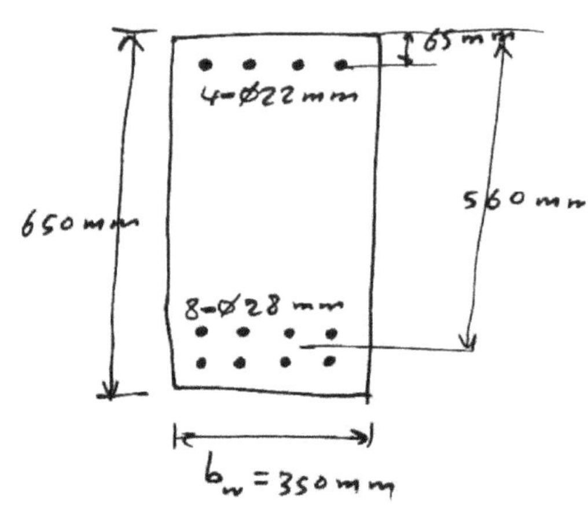

(a) Check whether tension steel exceeds the maximum $\boxed{14}$ amount permitted.

$$A_s' = 4\left[\frac{\pi(22)^2}{4}\right] = 1521 \text{ mm}^2.$$

$$\varepsilon_y = \frac{f_y}{E_s} = \frac{415}{200,000} = 0.002075$$

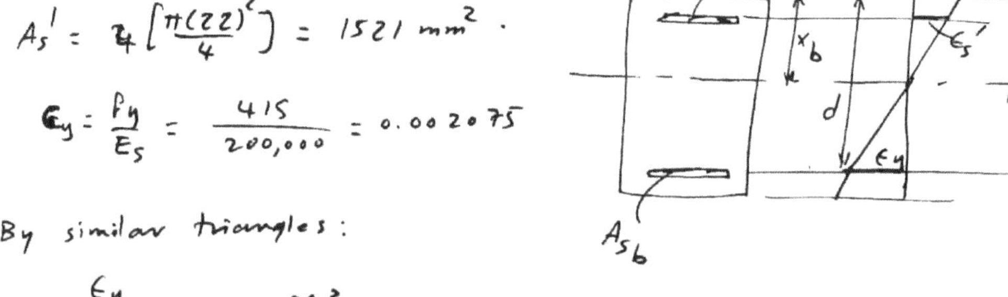

By similar triangles:

$$\frac{\varepsilon_y}{d - x_b} = \frac{0.003}{x_b}$$

$$\frac{0.002075}{560 - x_b} = \frac{0.003}{x_b} \implies 0.002075\, x_b = 0.003(560) - 0.003\, x_b$$

$$x_b = \frac{0.003(560)}{0.002075 + 0.003} = 331 \text{ mm}.$$

$$x_{max} = 0.75\, x_b$$
$$= 0.75(331)$$
$$= 248 \text{ mm}.$$

(doubly-reinforced beams)

For $x = 248$ mm:

$$\frac{\varepsilon_s'}{0.003} = \frac{248 - 65}{248}$$

$$\implies \varepsilon_s' = 0.00221 > \varepsilon_y = 0.002075$$

$$\implies \text{compression steel has yielded}$$

$$\implies f_s' = f_y.$$

$$C_{c_{max}} = 0.85 f_c'\, a_{max} b \qquad \text{where} \quad a_{max} = \beta_1 x_{max}$$

$\beta_1 = 0.85$ for $f_c' = 28$ MPa.

$$= 0.85 f_c'\,(0.85\, x_{max})\, b$$

$$= 0.85(28 \times 1000)(0.85)\left(\frac{248}{1000}\right)\left(\frac{350}{1000}\right)$$

$$= 1756 \text{ kN}.$$

$$C_{s_{max}} = (f_y - 0.85 f_c')A_s'$$

$$= \left[415 - 0.85(28)\right] \times 1000 \times 1521 \times 10^{-6}$$

$$= 595 \text{ kN}.$$

$$T_{max} = C_{c_{max}} + C_{s_{max}} = 1756 + 595 = 2351 \text{ kN}.$$

$$T_{max} = f_y A_{s\,max}$$

$$2351 = (415 \times 1000) A_{s\,max} \implies A_{s\,max} = 0.005665 \ m^2$$
$$= \underline{\underline{5663}} \ mm^2 .$$

$$\text{Used} \quad A_s = 8\left[\frac{\pi(28)^2}{4}\right] = 4926 \ mm^2 < A_{s\,max} = 5665 \ mm^2 .$$
$$\underline{\underline{o.k.}} \quad \smile$$

(b) Find the nominal flexural strength M_n:

<u>Assume compression steel yields:</u>

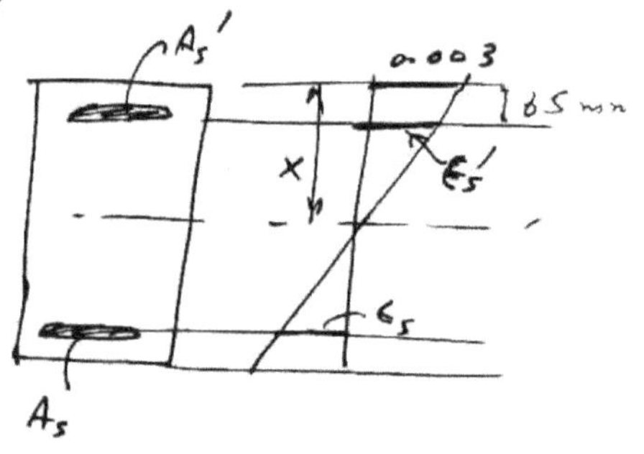

$$C_c + C_s = T$$

$$0.85 f_c' a b + (f_y - 0.85 f_c') A_s' = f_y A_s$$

$$\text{but} \quad a = \beta_1 x = 0.85 x$$

$$0.85(28)(0.85x)(350) + [415 - 0.85(28)](1521) = (415)(4926)$$

$$\therefore \quad x = 205 \ mm .$$

$$\frac{\epsilon_s'}{0.003} = \frac{205 - 65}{205} \implies \epsilon_s' = 0.00205$$
$$< \epsilon_y = 0.002075$$

$$\implies \text{compression steel does not yield:}$$

$$\implies \text{use } f_s' \text{ instead of } f_y:$$
and find new x

$$f_s' = E_s \epsilon_s'$$

$$\text{where} \quad \frac{\epsilon_s'}{0.003} = \frac{x - 65}{x}$$

$$\implies \epsilon_s' = 0.003\left(\frac{x-65}{x}\right)$$

$$\implies f_s' = 200,000(0.003)\left(\frac{x-65}{x}\right)$$

$$\implies f_s' = 600\left(\frac{x-65}{x}\right)$$

$$\therefore \quad C_c + C_s = T$$

$$0.85 f_c'(\beta_1 x) b + (f_s' - 0.85 f_c') A_s' = f_y A_s$$

$$0.85(28)(0.85)x(350) + \left[600\left(\frac{x-65}{x}\right) - 0.85(28)\right](1521) = (415)(4926)$$

$$\Rightarrow \quad 7080.5\,x^2 + 600(1521)(x-65) - 0.85(28)(1521)x = (415)(4926)x$$

$$7080.5\,x^2 - 2{,}080{,}489.8\,x + 912{,}600\,x - 600(1521)(65) = 0$$

$$7080.5\,x^2 - 1167889.8\,x - 600(1521)(65) = 0$$

$$\Rightarrow x = \frac{1167889.8 \pm \sqrt{(1167889.8)^2 - 4(7080.5)(-600)(1521)(65)}}{2(7080.5)}$$

$$= 206\ mm \cdot \quad \Rightarrow \quad a = \beta_1 x = 175\ mm \cdot$$

$$\Rightarrow \quad \varepsilon_s' = 0.003\left(\frac{206-65}{206}\right) = 0.00205 \quad < \varepsilon_y = 0.002073$$

$$\Rightarrow \quad f_s' = E_s \varepsilon_s' = 600\left(\frac{206-65}{206}\right) = 411\ mPa \cdot$$

$$\Rightarrow \quad \text{compression steel does not yield.}$$

$$C_c = 0.85 f_c' a b = 0.85 f_c'(\beta_1 x) b = 0.85(28 \times 1000)(0.85)\left(\frac{206}{1000}\right)\left(\frac{350}{1000}\right)$$

$$= 1458\ kN \cdot$$

$$C_s = (f_s' - 0.85 f_c')A_s'$$

$$= \left[411 - 0.85(28)\right] \times 1000 \times 1521 \times 10^{-6}$$

$$= 589\ kN \cdot$$

$$T = f_y A_s = (415 \times 1000) \times 4926 \times 10^{-6}$$

$$= 2044\ kN \cdot$$

$$\underline{\text{Check}}: \quad C_c + C_s = 1458 + 589 = 2047\ kN \simeq T \cdot$$

$$M_n = C_c\left(d - \frac{a}{2}\right) + C_s(d - d')$$

$$= 1458\left[\frac{560}{1000} - \frac{1}{2}\frac{175}{1000}\right] + 589\left[\frac{560}{1000} - \frac{65}{1000}\right]$$

$$= 980\ kN \cdot m \cdot$$

$$M_u \leq \phi M_n = 0.90(980) = 882 \text{ kN.m.}$$

but $\quad M_u = \dfrac{w_u L^2}{8}$

$$882 = \dfrac{w_u(7.3)^2}{8} \implies w_u \leq 132 \text{ kN/m.}$$

But $\quad \dfrac{w_L}{w_D} = 1.5 \implies w_L = 1.5\, w_D$

$$w_u = 1.4\, w_D + 1.7\, w_L$$
$$132 = 1.4\, w_D + 1.7(1.5\, w_D) = 3.95\, w_D$$
$$\implies w_D = 33.4 \text{ kN/m.}$$
$$w_L = 1.5(33.4) = 50.1 \text{ kN/m.}$$

(c) Design of Shear Reinforcement, using the simplified method with a constant value for V_c.

$\Rightarrow$ an envelope of maximum shear is needed.

$\Rightarrow$ the **live load** should always be treated as **variable** position loading acting wherever it may cause the greatest effect.

$\Rightarrow$ A **conservative solution**: an approximate shear envelope may be acceptable using a straight line relationship between the maximum shear at the support and the maximum shear at midspan!

Case (1): $\quad w_L$ is on whole span.

$V_{max} = \dfrac{w_u L}{2}$ at the centerline of the support.

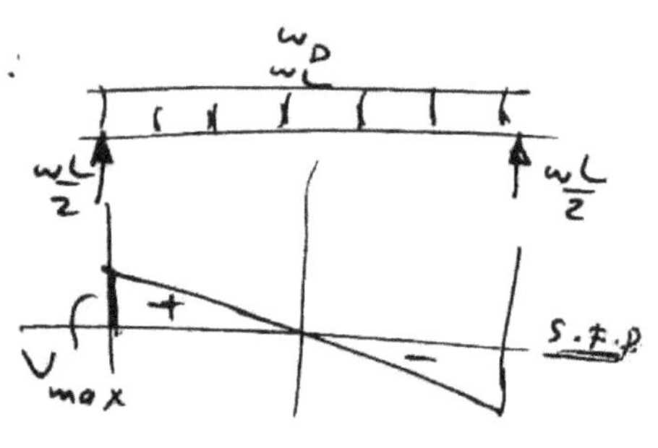

$$\therefore V_u = \dfrac{w_u L}{2} = \dfrac{\left[1.4(33.4) + 1.7(50.1)\right](7.3)}{2}$$

$$= 482 \text{ kN.}$$

Case 2 w_L is on half the span

At midspan:

$$V_u = \frac{w_L L}{8} = \frac{(1.7)(50.1)(7.3)}{8}$$

$$= 77.7 \text{ kN}$$

ACI Code / Section 11.1.3.1

Take the critical section at a distance d from the face of support,

    i.e. at $d = 560 \text{ mm}$.

$$\text{Support width} = \frac{7.3 - 7.0}{2} \times 2$$

$$= 0.3 \text{ m}$$

$$= 300 \text{ mm}$$

∴ Critical section is at $560 + 150 = 710 \text{ mm}$ from the center of the support

⟹ Determine V_u at the critical section.

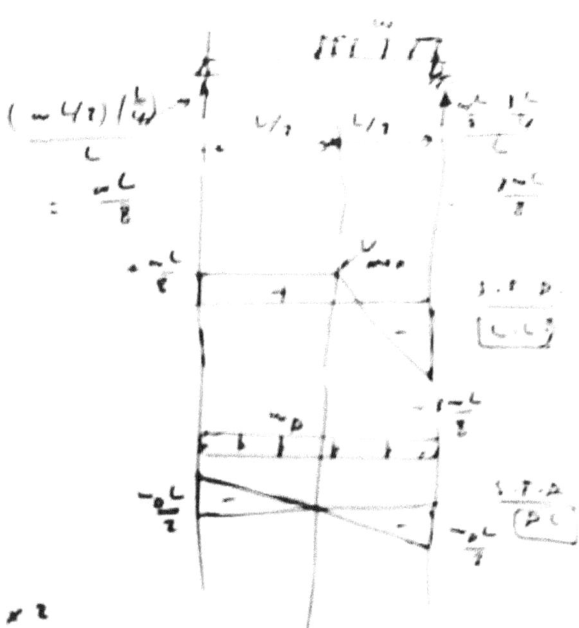

Midspan

approximate maximum shear envelope

true envelope

$+482 \text{ kN}$

$+727 \text{ kN}$

-77.7 kN

-482 k

7.3 m

℄ of support

critical section

7 m

7.3 m

0.3 m

Use linear interpolation : (at the critical section) :

$$V_u = 482 - \left(\frac{482 - 77.7}{\frac{1}{2}(7.3)}\right)\left(\frac{710}{1000}\right) = 403 \text{ kN}.$$

* the design requirement between the face of support and the critical section is considered to be constant $\equiv 403$ kN

Design Shear Strength Due to Concrete $\equiv \phi V_c$:

$$\phi V_c = \phi\left(\frac{1}{6}\sqrt{f_c'}\ b_w d\right) = 0.85\left[\frac{1}{6}\sqrt{28} \cdot (350)(560)\right]$$
$$= 147 \times 10^3 \quad N$$
$$= 147 \text{ kN}.$$

$$V_u \le \phi V_n = \phi(V_c + V_s) = \phi V_c + \phi V_s$$

$$\Longrightarrow \quad V_u \le \phi V_c + \phi V_s$$

Required $\boxed{\phi V_s = V_u - \phi V_c} = 403 - 147 = 256 \text{ kN}.$
(must be provided by shear reinforcement).

Check limits on ϕV_s :
$$\min \phi V_s = \phi \cdot \frac{1}{3} b_w d \qquad (\text{see page } \underline{135}).$$
$$= (0.85)\left(\frac{1}{3}\right)(350)(560)$$
$$= 55.5 \times 10^3 \, MN$$
$$= 55.5 \text{ kN}.$$

$$\therefore \quad \phi V_s > \min \phi V \qquad \underline{\underline{o \cdot k.}}$$

$$\max \phi V_s \le \phi\left(\frac{1}{3}\sqrt{f_c'}\ b_w d\right) \qquad \text{for} \quad s = d/2$$
$$\text{page 135} \qquad\qquad\qquad\qquad (ACI - 11.5.4.1$$
$$= 0.85\left(\frac{1}{3}\sqrt{28}\right)(350)(560) \qquad \& \ 11.5.4.3)$$
$$= 294 \times 10^3 \, MN$$
$$= 294 \text{ kN}.$$
$$\max \phi V_s \le \phi\left(\frac{2}{3}\sqrt{f_c'}\ b_w d\right) \qquad \text{for} \quad s = d/4$$
$$\qquad\qquad\qquad\qquad\qquad (ACI - 11.5.6.8$$
$$= (0.85)\left(\frac{2}{3}\sqrt{28}\right)(350)(560) \qquad \text{and} \ 11.5.4.3)$$
$$= 588 \times 10^3 \, MN$$
$$= 588 \text{ kN}.$$

Note: Since $\underbrace{\phi\left(\frac{1}{3}b_w d\right)}_{\min \phi V_s} <$ required $\phi V_s < \phi\left(\frac{1}{3}\sqrt{f_c'}\,b_w d\right)$

$$55.5 < 256\ kN < 294\ kN$$

$\Rightarrow$ design for the portion of the beam between the face of support and the critical section is in

<u>category 4</u> $\Rightarrow$ max. $s = \dfrac{d}{2} = \dfrac{560}{2} = 280\ mm < 600\ mm$

$\Rightarrow$ strength requirement using ϕV_s will control.

Using $\phi 10\ mm$ stirrups, we have: $\left[A_v = 2\left[\dfrac{\pi(10)^2}{4}\right] = 157\ mm^2\right]$

At the critical section, $s = \dfrac{\phi A_v f_y d}{\phi V_s} = \dfrac{(0.85)(157/1000)(415\times1000)\left(\frac{56}{100}\right)}{256}$

$$= 0.121\ m$$
$$= 121\ mm.$$

$\Rightarrow$ this means that $\phi 10\ mm$ stirrups may not be spaced farther apart than $121\ mm$ for the region from the face of support to the critical section at a distance d from the face of support.

$\Rightarrow$ A spacing of $120\ mm$ is acceptable.

$\Rightarrow$ For the region where minimum stirrups are required, the spacing required will be the lesser of $\frac{d}{2}$ ($280\ mm$)

and $\qquad s = \dfrac{3 A_v f_y}{b_w} \qquad$ (see page 134):

$$= \dfrac{3(157)\ (415)}{350} = 558\ mm.$$

$\Rightarrow$ required $s = 280\ mm.$

Note: As V_u decreases from the critical section toward midspan, the spacing s can be increased.

$\Rightarrow$ Prepare a table of suggested spacings.

z = distance from face of support to intersection of

$$\Phi V_n = (\Phi V_c + \Phi V_s) \quad \text{with} \quad V_u .$$

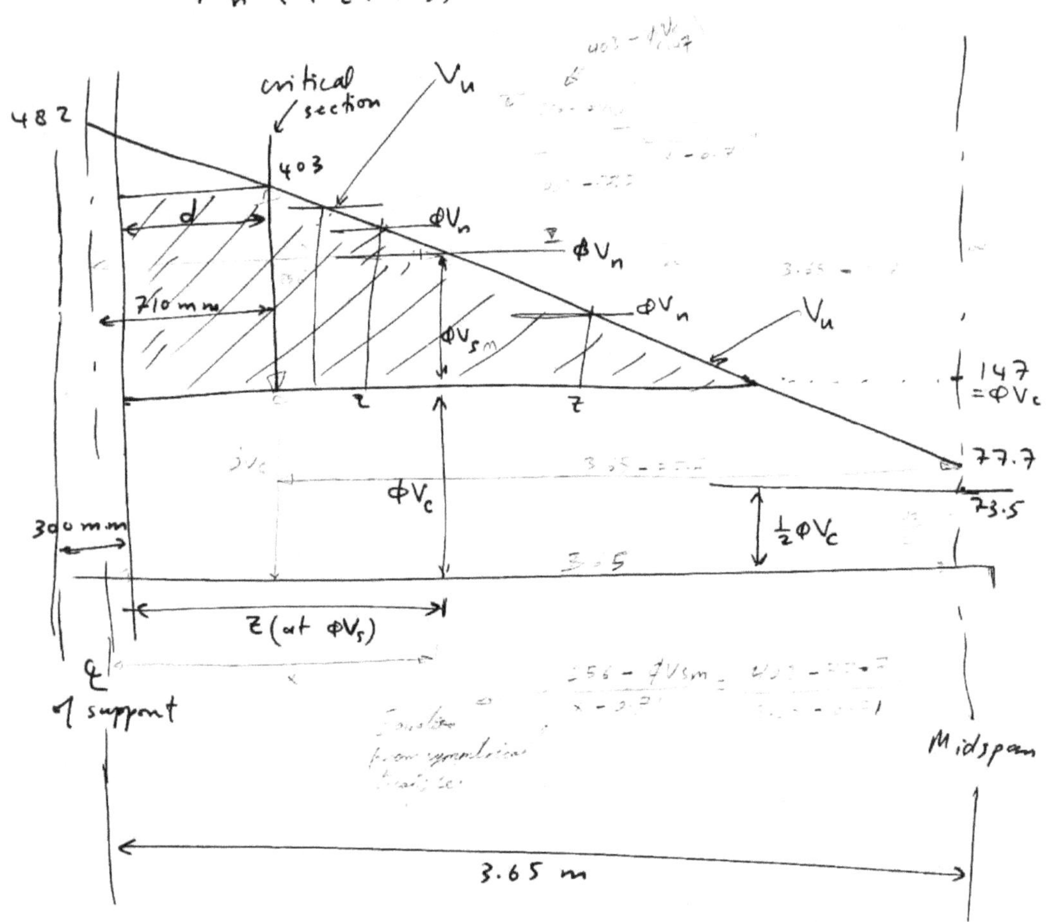

Using the Diagram :

1. Choose a spacing between the two limits (i.e. between 280 mm and 120 mm).

2. calculate the required ΦV_s.

3. calculate required $\Phi V_n = \Phi V_c + \Phi V_s$.

4. locate the intersection of ΦV_n with V_u on the diagram and scale off the value of z on the horizontal axis.

Using a Table :

By similar triangles:

$$\frac{256 - \Phi V_s}{403 - 77.7} = \frac{x - 0.710}{3.65 - 0.710}$$

$$\therefore \quad x - 0.710 = \frac{(256 - \Phi V_s)}{(403 - 77.7)}(3.65 - 0.710)$$

$$\therefore \quad x = 0.710 + \left(\frac{256 - \Phi V_s}{403 - 77.7}\right)(3.65 - 0.710)$$

But $z = x - 0.150$ (from the face of support)

$$\therefore \quad z = 0.56 + \left(\frac{256 - \Phi V_s}{403 - 77.7}\right)(3.65 - 0.710)$$

In mm,

$$z = 560 + \left(\frac{256 - \Phi V_s}{403 - 77.7}\right)(3650 - 710)$$

$\therefore$ Calculate z from the above formula instead of scaling from the diagram.

Choose the following intermediate spacings between the two extremes (120 mm and 280 mm):

Choose $s = 120\,mm \rightarrow 145\,mm \rightarrow 170\,mm \rightarrow 225\,mm \rightarrow 280\,m$

Calculate $\Phi V_s = \dfrac{\Phi A_v f_y d}{s}$ for each case

$$= \frac{0.85\,(157 \times 10^{-6})\,(415 \times 1000)\left(\frac{560}{1000}\right)}{s}$$

$$\Rightarrow \Phi V_s = \frac{31}{s} \quad (s \text{ in } m).$$

S (mm)	ΦV_s (kN)	z (mm) (from the formula above)
121 mm	256 (max.)	0 to 560 mm.
145	214	940 mm
170	182	1229 mm
225	138	1626 mm
280 (max.)	111	1870 mm
559 (not used)	55.5 (min)	2372 mm.

Place the first stirrup at 100 mm from the face of support : $\lfloor \tfrac{150}{\text{actual } z}$

no stirrup $\longrightarrow$ 100 mm.

8 @ 120 mm $\longrightarrow$ 100 + 840 = 940 mm.

3 @ 145 mm $\longrightarrow$ 940 + 290 = 1230 mm.

4 @ 170 mm $\longrightarrow$ 1230 + 510 = 1740 mm.

7 @ 280 mm $\longrightarrow$ 1740 + 1680 = 3420 mm.

∴ provide ϕ10 mm stirrups spaced as follows :

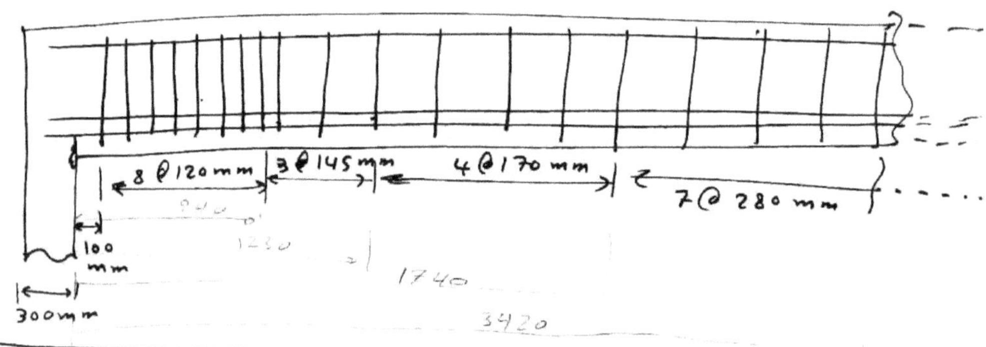

Example 2 :

For the beam designed in Example 1, investigate whether or not the designer could justify increasing the 120-mm spacing used near the support to 150 mm by applying the more detailed procedure for V_c involving $\dfrac{\rho_w V_u d}{M_u}$ (ACI – 11.3.2.1).

Solution :

At the critical section :

$$V_u = 403 \text{ kN}, \qquad A_s = 8\left[\frac{\pi(28)^2}{4}\right] = 4926 \text{ mm}^2.$$

$$V_c = \frac{1}{6}\left(\sqrt{f_c'} + 100\, \frac{\rho_w V_u d}{M_u}\right) b_w d \;\leq\; 0.3\sqrt{f_c'}\, b_w d$$

$$\text{and} \quad \frac{V_u d}{M_u} \leq 1.0 \qquad (ACI\ 11.3.2.1).$$

$$\rho_w = \frac{A_s}{b_w d} = \frac{4926}{350 \times 560} = 0.0251$$

$$M_u = \frac{w_u L}{2}(0.710) - w_u \frac{(0.710)^2}{2}$$

$$= \frac{w_u(0.710)}{2}\left(L - 0.710\right)$$

$$= \frac{(132)(0.710)}{2}\left(7.3 - 0.710\right)$$

$$= 309 \text{ kN} \cdot \text{m}.$$

$$\frac{V_u d}{M_u} = \frac{(403)(0.560)}{309} = 0.730 \;<\; 1.0 \quad \underline{o.k.}$$

$$\therefore V_c = \frac{1}{6}\left[\sqrt{28} + 100\,(0.0251)(0.730)\right](350)(560)$$

$$= 232711 \text{ N}$$

$$\cong 233 \text{ kN}.$$

Check that $V_c \leq 0.3 \sqrt{f_c'} \, b_w d$

$$0.3 \sqrt{f_c'} \, b_w d = 0.3 \sqrt{28} \, (350)(560) = 311140 \text{ N}$$
$$= 311 \text{ kN}.$$

$\therefore V_c = 233 \text{ kN} < 311 \text{ kN}. \qquad \underline{\text{o.k}} \; \smile$

$\therefore \phi V_c = (0.85)(233) = \underline{198 \text{ kN}}.$

Compare with $\phi V_c = 147 \text{ kN}$ of the simplified procedure. This gives more concrete strength, thus reducing shear reinforcement needed.

* Stirrup spacing for $\phi 10 \text{mm}$ stirrups:

At the critical section:

$$V_u \leq \phi V_n = \phi (V_c + V_s) = \phi V_c + \phi V_s$$

$$\phi V_s \geq V_u - \phi V_c = 403 - 198 = 205 \text{ kN}.$$

$$\therefore \text{max. } S = \frac{\phi A_v f_y d}{\phi V_s} = \frac{(0.85)\left(2\left(\frac{(10)^2}{4}\right) \times 10^{-6} \times (415 \times 1000)\right)\left(\frac{560}{1000}\right)}{205}$$

$$= 0.151 \text{ m}$$
$$= 151 \text{ mm}.$$

$\qquad \rightarrow$ Take max. $S = 150 \text{ mm}.$

$\therefore$ Increasing the stirrup spacing from 120 mm to 150 mm is justified using the detailed procedure for calculating V_c.

<u>Web Reinforcement — Types of Stirrups:</u>

* <u>Shear reinforcement</u> (e.g. stirrups) is also called <u>web reinforcement</u>

* The most common stirrups are U-shaped, but they can be ⊔⊔-shaped, or perhaps having only a single vertical leg.

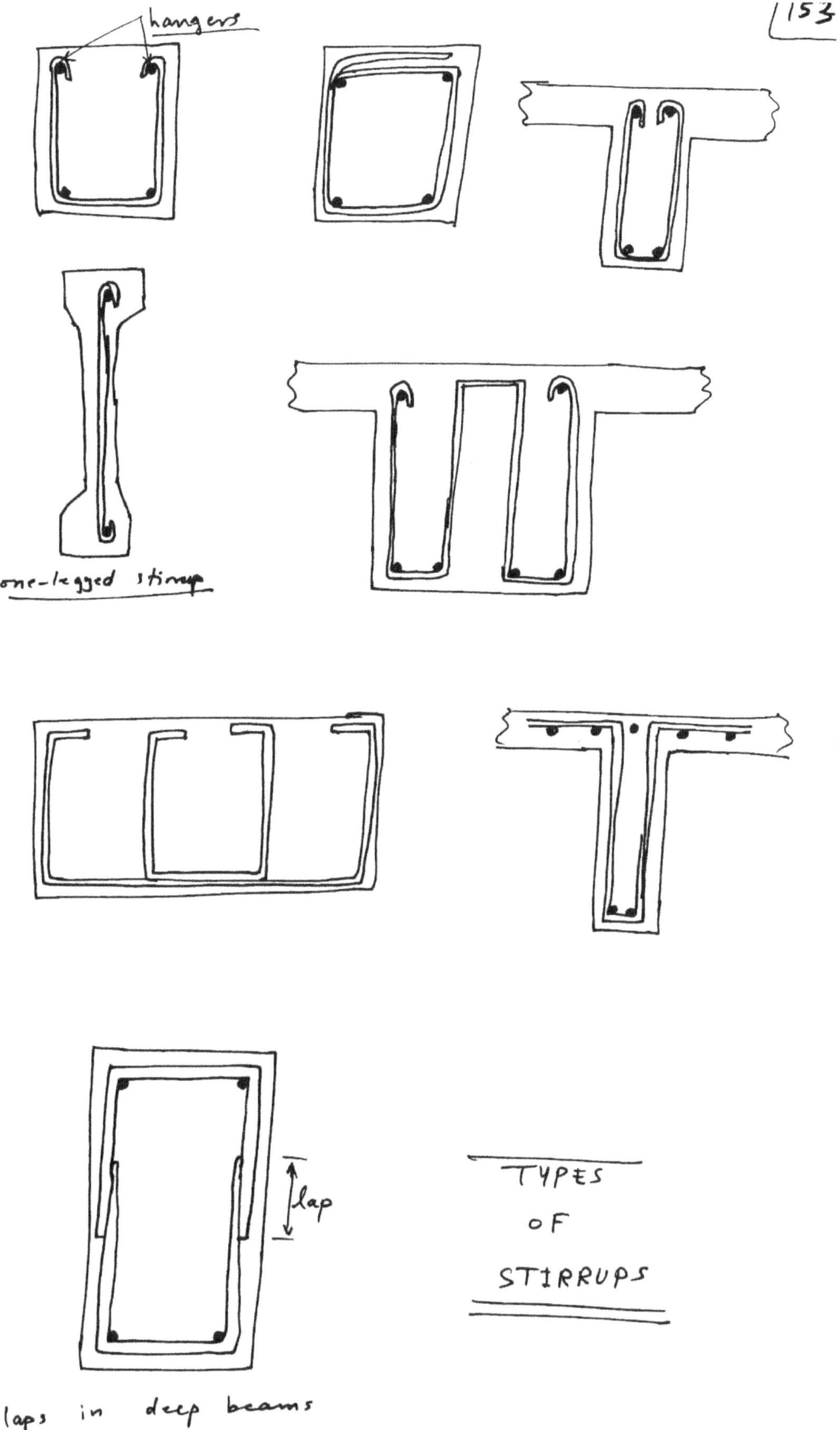

hangers
one-legged stirrup
laps in deep beams
TYPES
OF
STIRRUPS
lap

Example 3:

Determine the minimum cross-section required for a rectangular beam from a shear standpoint so that <u>no</u> web reinforcing is required by the ACI Code if

$$V_u = 170 \text{ kN} \quad \text{and} \quad f_c' = 28 \text{ MPa}.$$ Use the conservative value of V_c (simplified procedure).

Solution:

Shear strength provided by concrete is given by the following conservative equation (simplified procedure):

$$\phi V_c = \phi \left(\tfrac{1}{6} \sqrt{f_c'} \, b_w d \right) = \phi \left(\tfrac{1}{6} \sqrt{28} \, b_w d \right)$$

$$= 0.85 \left(\tfrac{1}{6} \sqrt{28} \right) b_w d \qquad , \quad N$$

$$= \frac{0.85}{1000} \left(\tfrac{1}{6} \sqrt{28} \right) b_w d \qquad , \quad kN$$

$$= 0.00074963 \, b_w d \qquad , \quad kN.$$

if $\qquad V_u \leq 0.5 \phi V_c \qquad$ (no shear reinforcement is required / ACI Code / Section 11.5.5.1)

$\therefore$ let $\qquad 170 \leq 0.5 \, (0.00074963 \, b_w d)$

$$\Rightarrow \quad b_w d \geq 453,557 \text{ mm}^2.$$

let $\quad b_w = 550 \text{ mm} \quad \Rightarrow \quad d \geq 825 \text{ mm}.$

Use a 550×900 mm section.
(no stirrups are required.

---

* <u>Minimum Spacing of Stirrups:</u>

*Stirrups must be placed far enough apart to permit the aggregate to pass through, and, in addition, they must be reasonably few in number so as to keep within reason the amount of labor involved in placing them.

* Minimum spacings of 75 mm or 100 mm are normally used.
* Usually $\phi 10$ mm stirrups are assumed.
* If calculated design spacings are less than $\frac{d}{4}$, larger-diameter stirrups can be used.
* Another alternative is to use ⊔⊔-stirrups instead of U-stirrups.
* Different diameter stirrups should not be used in the same beam or confusion will result.

Example 4 :

The beam shown in the figure was selected using $f_y = 415$ MPa and $f_c' = 20$ MPa. Determine the theoretical spacing of $\phi 10$ mm U-stirrups for each of the following shears:

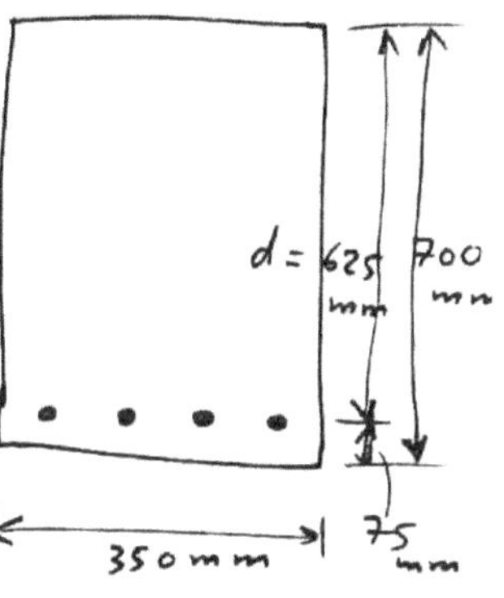

    (a) $V_u = 62$ kN.
    (b) $V_u = 245$ kN.
    (c) $V_u = 310$ kN.
    (d) $V_u = 715$ kN.

Solution :

(a) $V_u = 62$ kN.

$$\phi V_c = \phi\left(\frac{1}{6}\sqrt{f_c'}\, b_w d\right) = (0.85)\left(\frac{1}{6}\sqrt{20}\right)(350)(625)/1000$$
$$= 139 \text{ kN.}$$

$$0.5\,\phi V_c = \frac{139}{2} = 69.5 \text{ kN.}$$

$$\therefore \quad V_u \leq 0.5\,\phi V_c \qquad \text{since } 62 < 69.5 \text{ kN.} \implies \boxed{\text{Category } ①}$$

$\therefore$ Stirrups are <u>not</u> required.

(b) $V_u = 245 \text{ kN}$.

$$V_u > 0.5 \, \phi V_c = 69.5 \text{ kN}.$$

$\therefore$ stirrups are needed.

Also, $V_u > \phi V_c = 139 \text{ kN}$.

$\therefore$ This is neither categories ① or ②.

Check minimum shear reinforcement:

$$\min \phi V_s = \phi \left(\tfrac{1}{3} b_w d\right) = 0.85 \left(\tfrac{1}{3}\right)(330)(625)/1000$$

$$= 62 \text{ kN}.$$

$\therefore \phi V_c + \min \phi V_s = 139 + 62 = 201 \text{ kN}.$

$$V_u > \left[\phi V_c + \min \phi V_s\right].$$

$\therefore$ this is $\underline{not}$ category ③.

Calculate $\phi\left(\tfrac{1}{3}\sqrt{f_c'}\right) b_w d = 0.85\left(\tfrac{1}{3}\sqrt{20}\right)(350)(625)/1000$

$$= 277 \text{ kN}.$$

$\therefore \left[\phi V_c + \phi\left(\tfrac{1}{3}\sqrt{f_c'}\right)b_w d\right] = 139 + 277 = 416 \text{ kN}.$

$$\left[\phi V_c + \min \phi V_s\right] < V_u \leq \left[\phi V_c + \phi\left(\tfrac{1}{3}\sqrt{f_c'}\right)b_w d\right]$$

$\therefore$ this is $\boxed{\text{Category ④}}$.

$\therefore$ required $\phi V_s = V_u - \phi V_c = 245 - 139 = 106 \text{ kN}.$

$$\therefore s \leq \frac{\phi A_v f_y d}{\phi V_s} = \frac{(0.85)\left[2 \cdot \frac{\pi (10)^2}{4} \times 10^{-6}\right](415\times1000)\left(\frac{625}{1000}\right)}{106}$$

$$= 0.327 \text{ m}$$

$$= 327 \text{ mm}.$$

$\text{Max. } s = \frac{d}{2} \leq 600 \text{ mm}.$

$\therefore \frac{d}{2} = \frac{625}{2} = 312.5 \text{ mm}.$

$\therefore$ Take $\underline{\underline{s = 312 \text{ mm}}}$.

(c) $V_u = 310$ kN.

$$\left[\phi V_c + \min \phi V_s\right] < V_u \le \left[\phi V_c + \phi\left(\tfrac{1}{3}\sqrt{f_c'}\right)b_w d\right]$$

$\Rightarrow$ this is $\boxed{\text{Category } \textcircled{4}}$.

$\therefore$ required $\phi V_s = V_u - \phi V_c = 310 - 139 = 171$ kN.

$$s \le \frac{\phi A_v f_y d}{\phi V_s} = \frac{(0.85)\left[2\left(\frac{\pi (10)^2}{4}\right)\times 10^{-6}\right](415\times 1000)\left(\frac{625}{1000}\right)}{171}$$

$$= 0.202 \text{ m}$$

$$= 202 \text{ mm}.$$

max. $s = \dfrac{d}{2} = \dfrac{625}{2} = 312.5$ mm ≤ 600 mm.

$\therefore$ Take $\underline{s = 202 \text{ mm}}$.

(d) $V_u = 715$ kN.

Obviously, $\qquad V_u > \left[\phi V_c + \phi\left(\tfrac{1}{3}\sqrt{f_c'}\right)b_w d\right]$.

Calculate $\quad \phi\left(\tfrac{2}{3}\sqrt{f_c'}\right)b_w d = 0.85\left(\tfrac{2}{3}\sqrt{20}\right)(350)(625)/1000$

$$= 554 \text{ kN}.$$

$\therefore \left[\phi V_c + \phi\left(\tfrac{2}{3}\sqrt{f_c'}\right)b_w d\right] = 139 + 554 = 693$ kN.

$\therefore \qquad V_u > \left[\phi V_c + \phi\left(\tfrac{2}{3}\sqrt{f_c'}\right)b_w d\right]$.

this is $\underline{not}$ even Category $\textcircled{5}$.

$\Rightarrow$ this case does $\underline{not}$ fall in any of the five categories

$\Rightarrow$ We need a larger beam.

Re-design the beam dimensions.

Example 5 :

Select $\phi 10\,mm$ U-stirrups for the beam shown in the figure, for which $w_D = 60\,kN/m$ and $w_L = 90\,kN/m$.
Take $f_c' = 28\,MPa$ and $f_y = 415\,MPa$.

Solution :

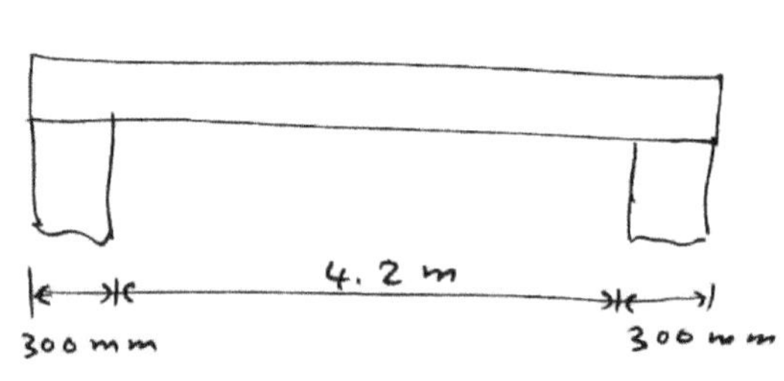

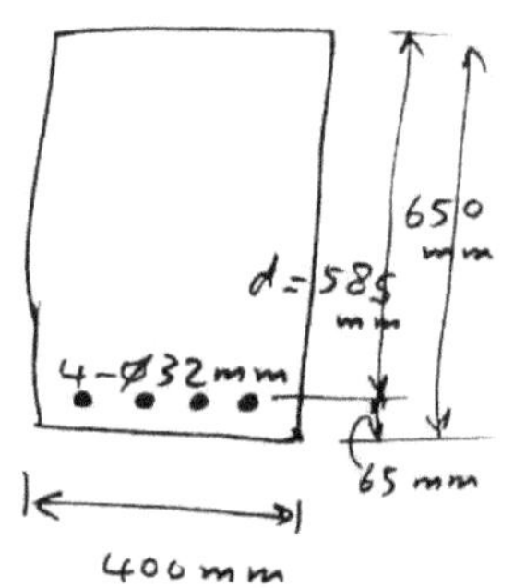

$$w_u = 1.4\,w_D + 1.7\,w_L = 1.4(60) + 1.7(90) = 237\,kN/m .$$

Note : In this example, we will assume that the live load covers the whole span.

⇒ there is only one case.

⇒ We only need the shear diagram.

At the left end (centerline of support):

$$V_u = \frac{w_u L}{2} = \frac{237(4.5)}{2} = 533.25\,kN .$$

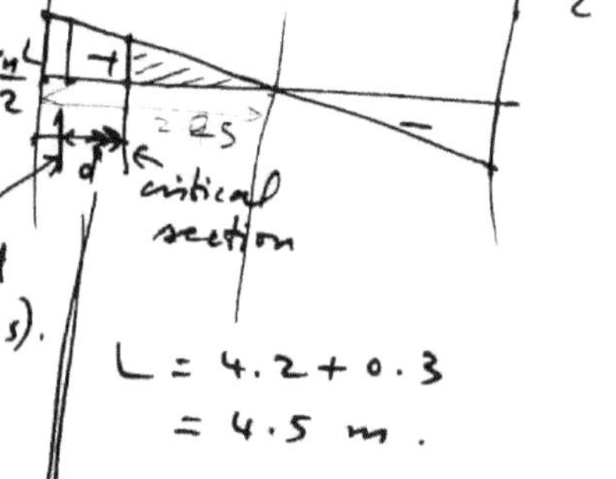

At the face of the support :

$$V_u = \left(\frac{\left(\frac{4.2}{2}\right)}{\left(\frac{4.5}{2}\right)}\right)(533.25) \quad \text{(similar triangles)}.$$

$$= 497.7\,kN$$

At the critical section (distance d from the face of support):

$$V_u = \frac{\left(\frac{4.2}{2} - 0.585\right)}{\left(\frac{4.5}{2}\right)}(533.25) = 359\,kN . \quad \text{(similar triangles)}$$

Using the simplified procedure :

$$\phi V_c = \phi\left(\frac{1}{6}\sqrt{f_c'}\right)b_w d = (0.85)\left(\frac{1}{6}\sqrt{28}\right)(400)(585)/1000 = 175\,kN .$$

Since $V_u > \phi V_c \Rightarrow$ this is neither Category $\boxed{159}$
        (1) or (2).

$$\min \phi V_s = \phi\left(\tfrac{1}{3} b_w d\right) = 0.85 \left(\tfrac{1}{3}\right)(400)\frac{(585)}{1000} = 66.3 \text{ kN}.$$

$$\phi V_c + \min \phi V_s = 175 + 66.3 = 241.3 \text{ kN}.$$

But $\quad V_u = 359 \text{ kN} > \left[\phi V_c + \min \phi V_s\right] = 241.3 \text{ kN}.$

$\therefore$ this is not Category (3).

$\therefore$ Calculate $\quad \phi\left(\tfrac{1}{3}\sqrt{f_c'}\right) b_w d = 0.85\left(\tfrac{1}{3}\sqrt{28}\right)(400)\frac{(585)}{1000} = 351 \text{ kN}.$

$$\therefore \left[\phi V_c + \phi\left(\tfrac{1}{3}\sqrt{f_c'}\right) b_w d\right] = 175 + 351 = 526 \text{ kN}.$$

$$\therefore \left[\phi V_c + \min \phi V_s\right] < V_u \le \left[\phi V_c + \phi\left(\tfrac{1}{3}\sqrt{f_c'}\right) b_w d\right]$$

$\Rightarrow$ this is Category (4).

$\therefore$ required $\phi V_s = V_u - \phi V_c = 359 - 175 = 184 \text{ kN}.$

$$\therefore \quad \text{max. } s = \frac{\phi A_v f_y d}{\phi V_s}$$

$$= \frac{(0.85)(2)\left[\frac{\pi(10)^2}{4}\right]\times 10^{-6}(415\times1000)\left(\frac{585}{1000}\right)}{184}$$

$$= 0.176 \text{ m}$$

$$= 176 \text{ mm}.$$

Also check $\quad s \le \dfrac{d}{2} \le 600 \text{ mm}.$

$$\uparrow \frac{d}{2} = \frac{585}{2} = 292.5 \text{ mm} \quad \longleftarrow \underline{\underline{\text{o.k}}}$$

$\Rightarrow$ Take $\underline{\underline{s = 176 \text{ mm}}}.$

For the region where minimum stirrups are required,
the spacing required will be the lesser of $\frac{d}{2} = 292 \text{ mm}$

and $\quad s = \dfrac{3 A_v f_y}{b_w} = \dfrac{3\left[2\frac{\pi(10)^2}{4}\right](415)}{400} = 489 \text{ mm}.$

$\Rightarrow$ Take $\underline{s = 292 \text{ mm}}.$

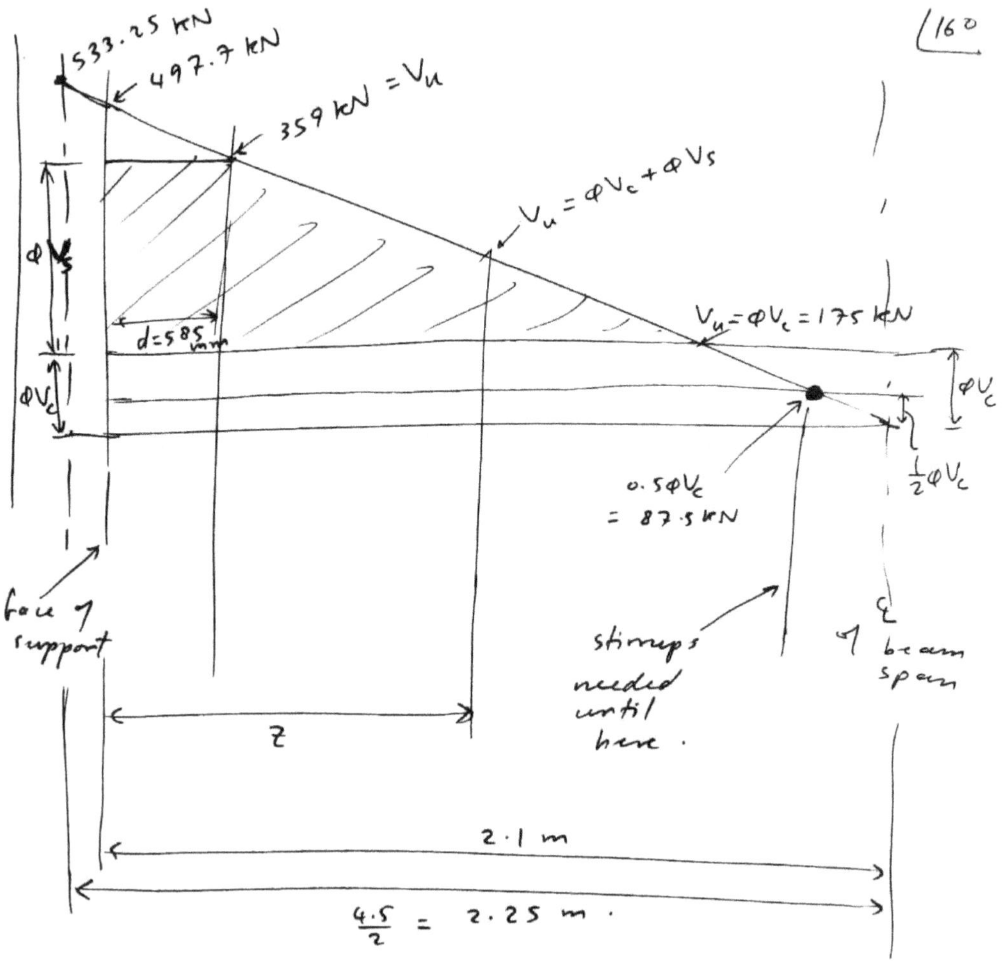

Z = distance from face of support to intersection of
V_u with $\phi V_n = \phi V_c + \phi V_s$.

By similar triangles:

$$\frac{184 - \phi V_s}{359 - 0} = \frac{x - 0.735}{2.25 - 0.735}$$

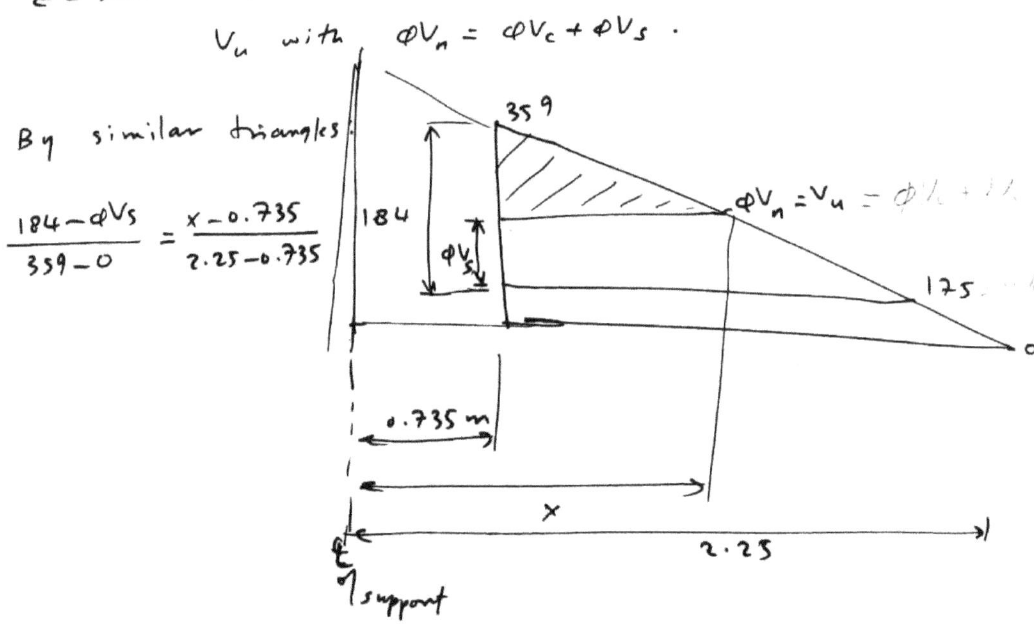

$$\therefore \quad x - 0.735 = \frac{184 - \phi V_s}{359}(2.25 - 0.735)$$

$$x = 0.735 + \left(\frac{184 - \phi V_s}{237}\right)$$

but $\quad z = x - 0.150$

$$\Rightarrow \boxed{z = 0.585 + \left(\frac{184 - \phi V_s}{237}\right)} \quad \text{\textcircled{d}}$$

$\therefore$ Calculate z from the above formula instead of scaling from the diagram.

Choose the following intermediate spacings between the two extremes (176 mm and 292 mm).

Choose $s = 175$ mm $\rightarrow 250$ mm $\rightarrow 290$ mm.

Calculate $\phi V_s = \dfrac{\phi A_v f_y d}{s}$ for each case

$$\therefore \quad \phi V_s = \frac{0.85\,(157 \times 10^{-6})(415 \times 1000)\left(\frac{585}{1000}\right)}{s}$$

s (mm)	ϕV_s (kN)	z (mm) (from above formula)
176 mm	184 (max.)	0 to 585 mm.
250 mm	130	812 mm.
292 mm (max.)	111	893 mm.
489 mm (not used)	66.3 (min.)	1082 mm.

At $\phi V_n = V_u = 0.5\,\phi V_c = \dfrac{175}{2} = 87.5$ kN.

(no stirrups are needed beyond this point).

i.e. at $\quad z = 2.1 - \left(\dfrac{87.5}{497.7}\right)(2.1) = 1.73$ m.

Place the first stirrup at 100 mm from the face of support:

$$
\begin{aligned}
\text{no stirrup} &\longrightarrow 100 \text{ mm} \\
5 @ 175 \text{ mm} &\longrightarrow 100 + 700 = 800 \text{ mm} \\
1 @ 250 \text{ mm} &\longrightarrow 800 + 250 = 1050 \text{ mm} \\
4 @ 290 \text{ mm} &\longrightarrow 1050 + 870 = 1920 \text{ mm} \\
\text{no stirrup} &\longrightarrow (1920 \text{ mm} \rightarrow 2100 \text{ mm})
\end{aligned}
$$

(symmetric about $\mathcal{C}$).

∴ Provide ⌀10 mm stirrups spaced as follows:

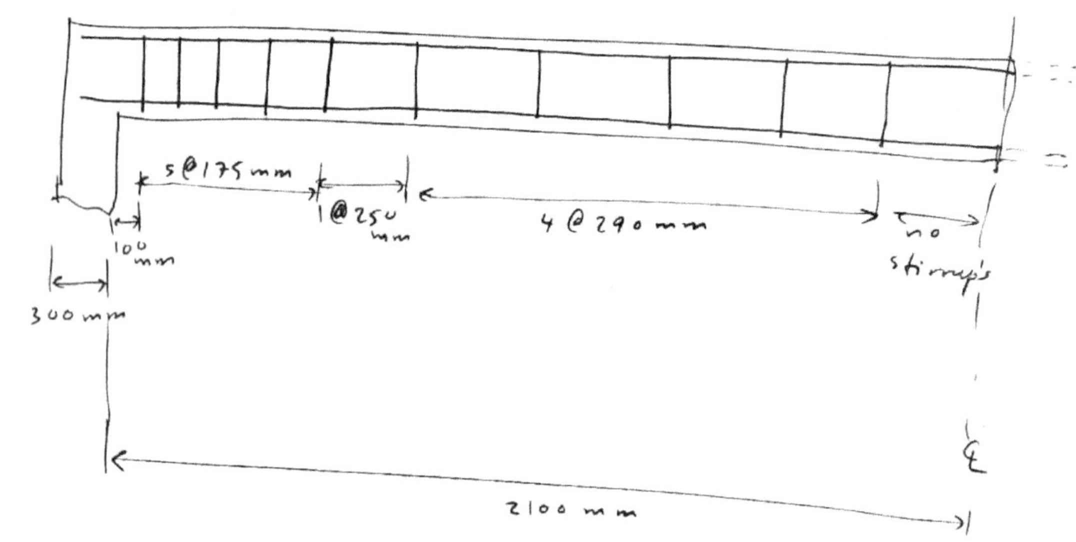

Economical Spacing of Stirrups:

* When stirrups are required in a reinforced concrete member, the Code specifies maximum permissible spacings varying from $\frac{d}{4}$ to $\frac{d}{2}$.

* Stirrup spacings less than $\frac{d}{4}$ are <u>not</u> economical.

* Many designers use a maximum of three different spacings in a beam. These are $\frac{d}{4}$, $\frac{d}{3}$, and $\frac{d}{2}$.

* It is easily possible to derive a value of ϕV_s for each size and style of stirrups for each of these spacings

Using $\phi 10\,mm$ stirrups and $f_y = 415\ MPa$,

$$A_v = 2\left[\frac{\pi (10)^2}{4}\right] = 157\ mm^2.$$

$$\phi V_s = \frac{\phi A_v f_y d}{s}$$

$$= \frac{(0.85)(157\times10^{-6})(415\times1000)\,d}{d/2}$$

$$= 111\ kN \qquad (\text{for } s = d/2).$$

$$\phi V_s = \frac{(0.85)(157\times10^{-6})(415\times1000)\,d}{d/3}$$

$$= 166\ kN \qquad (\text{for } s = d/3)$$

$$\phi V_s = \frac{(0.85)(157\times10^{-6})(415\times1000)\,d}{d/4}$$

$$= 222\ kN \qquad (\text{for } s = d/4).$$

$\phi 10\,mm$	$f_y = 415\ MPa$
s	ϕV_s
$d/2$	$111\ kN$
$d/3$	$166\ kN$
$d/4$	$222\ kN$

* Similar tables can be produced for different steels (f_y) and different stirrups (A_v) by applying the above formula.

5.14 Shear Strength of Members Under Combined Bending and Axial Load:

* For the same bending moment, the shear strength of a member is.
  (1) increased by adding an axial compressive load.
  (2) decreased by adding an axial tensile load.

* **Axial Compression :**

* the addition of axial compression tends to delay the opening of the shear crack and prevent its extending as far into the beam.

* For the simplified method with axial compression, a linear increase with axial compression is given by the ACI Code / Section 11.3.1.2 :

the previous equation $V_c = \frac{1}{6} \sqrt{f_c'} \, b_w d$

becomes $V_c = \left(1 + \frac{N_u}{14 A_g}\right) \frac{\sqrt{f_c'}}{6} b_w d$,

$\frac{N_u}{A_g}$ and f_c' in MPa.

where $N_u \equiv$ factored axial compressive force

$A_g \equiv$ gross-cross-sectional area of concrete.

* The more detailed procedure for calculating V_c may be used where :

$$V_c = \frac{1}{6}\left(\sqrt{f_c'} + 100 \rho_w \frac{V_u d}{M_u}\right) b_w d \leq 0.3 \sqrt{f_c'} \, b_w d$$

$$\underline{\underline{or}} \quad V_c = \frac{1}{7}\left(\sqrt{f_c'} + 120 \rho_w \frac{V_u d}{M_u}\right) b_w d \leq 0.3 \sqrt{f_c'} \, b_w d$$

with $\frac{V_u d}{M_u} \leq 1.0$

(ACI / Section 11.3.2.1).

In the use of the above equation for axial compression, M_m replaces M_u, where:

$$M_m = M_u - N_u \left(\frac{4h - d}{8} \right) \qquad \text{(ACI/Section 11.3.2.2)}$$

In this case, $\quad V_c \leq 0.3 \sqrt{f_c'}\, b_w d \sqrt{1 + \frac{0.3 N_u}{A_g}}$

(maximum value).

---

* __Axial Tension__ : For members subject to axial tension:

Simplified Procedure: $\qquad V_c = 0 \qquad$ (ACI/Section 11.3.1.3)

Detailed Procedure: $\qquad V_c = \left(1 + \frac{0.3 N_u}{A_g} \right) \frac{\sqrt{f_c'}}{6} b_w d$

(ACI/Section 11.3.2.3)

where N_u is <u>negative</u> for tension and $\frac{N_u}{A_g}$ and f_c' are in MPa.

<u>Note</u> : V_c in the above formula is <u>reduced</u>.

__Example 6__ :

Show the effect of axial load on the ACI shear strength V_c for the beam of Example 1 when it contains no shear reinforcement. Compute V_c for the critical section at d from the face of support, where $V_u = 403$ kN and $M_u = 369$ kN·m (computed in Example 2). As in Example 1, use $f_c' = 28$ MPa, $b = 350$ mm, $d = 560$ mm, $h = 650$ mm, and $8-\phi 28$ mm bars for the tension steel.

__Solution:__

(a) __Axial Compression__.

Simplified Procedure: $\quad V_c = \left(1 + \frac{N_u}{14 A_g} \right) \frac{\sqrt{f_c'}}{6} b_w d$

$$\Rightarrow \quad \frac{V_c}{\sqrt{f_c'}\,b_w d} = \frac{1}{6}\left(1 + \frac{N_u}{14 A_g}\right) = \frac{1}{6} + \frac{N_u}{84 A_g}$$

$\therefore \dfrac{V_c}{\sqrt{f_c'}\,b_w d}$ increases linearly with $\dfrac{N_u}{A_g}$.

at $\dfrac{N_u}{A_g} = 0 \quad \Longrightarrow \quad \dfrac{V_{c\,(min.)}}{\sqrt{f_c'}\,b_w d} = \dfrac{1}{6} = 0.167$

at $\dfrac{N_u}{A_g} = 5.5 \quad \Longrightarrow \quad \dfrac{V_{c\,(max.)}}{\sqrt{f_c'}\,b_w d} = 0.232$

More Detailed Procedure:

$$V_c = \frac{1}{6}\left(\sqrt{f_c'} + 100\,\rho_w \frac{V_u d}{M_u}\right) b_w d$$

$$\therefore \quad \rho_w = \frac{A_s}{bd} = \frac{8\left[\frac{\pi (28)^2}{4}\right]}{350 \times 560} = 0.0251$$

$$\therefore \quad \frac{V_c}{\sqrt{f_c'}\,b_w d} = \frac{1}{6}\left(1 + 100\,\rho_w \frac{V_u d}{M_u \sqrt{f_c'}}\right)$$

$$= \frac{1}{6} + \frac{100}{6}\,\rho_w \frac{V_u d}{M_u \sqrt{f_c'}}$$

$$= \frac{1}{6} + \frac{100}{6}(0.0251)\left(\frac{V_u}{M_u}\right)\left(\frac{560}{\sqrt{28}}\right)\Big/ 1000$$

$$\Rightarrow \quad \frac{V_c}{\sqrt{f_c'}\,b_w d} = \frac{1}{6} + \frac{44.27}{1000}\left(\frac{V_u}{M_u}\right)$$

But replace M_u by M_m which is given by:

$$M_m = M_u - N_u\left(\frac{4h - d}{8}\right)$$

$$= M_u - N_u\left[\frac{4(0.650) - 0.560}{8}\right]$$

$$\therefore M_m = M_u - 0.235\,N_u$$

At the critical section, $M_u = 309 \text{ kN·m}$ and

$V_u = 403 \text{ kN·m}$.

$\therefore \quad m_m = 309 - 0.255 N_u$, but $A_g = \dfrac{350 \times 560}{(1000)(1000)}$

$= 309 - 0.255 N_u \cdot \dfrac{1}{A_g}(0.350)(0.560)$

$\therefore \quad M_m = 309 - 0.05 \dfrac{N_u}{A_g}$, $\dfrac{N_u}{A_g}$ in MPa.

$\therefore \quad \dfrac{V_c}{\sqrt{f_c'}\, b_w d} = \dfrac{1}{6} + \dfrac{44.27}{1000}\left[\dfrac{403}{309 - 0.05 \frac{N_u}{A_g}}\right]$

at $\dfrac{N_u}{A_g} = 0 \implies \dfrac{V_{c\,(min.)}}{\sqrt{f_c'}\, b_w d} = 0.224$.

at $\dfrac{N_u}{A_g} = \underset{(\times 1000)}{5.5} \implies \dfrac{V_{c\,(max.)}}{\sqrt{f_c'}\, b_w d} = 0.691$.

(b) **Axial Tension** :

Simplified Procedure: $V_c = 0$.

Detailed Procedure: $V_c = \left(1 + 0.3 \dfrac{N_u}{A_g}\right)\dfrac{\sqrt{f_c'}}{6} b_w d$

where N_u is <u>negative</u>.

$\implies V_c$ is reduced by axial tension

$\therefore \quad \dfrac{V_c}{\sqrt{f_c'}\, b_w d} = \dfrac{1}{6}\left(1 + \dfrac{0.3 N_u}{A_g}\right)$.

check when V_c will become zero.

$\dfrac{1}{6}\left(1 + 0.3\dfrac{N_u}{A_g}\right) = 0$

$0.3\dfrac{N_u}{A_g} = -1 \implies \boxed{\dfrac{N_u}{A_g} = -3.33}$

$\therefore \quad$ At $\dfrac{N_u}{A_g} = 0 \implies \dfrac{V_{c\,(max.)}}{\sqrt{f_c'}\, b_w d} = \dfrac{1}{6}$

At $\dfrac{N_u}{A_g} = -3.33 \implies V_c = 0$

*** Supplement to T Beams :**

Effective Flange Width. ACI Code / Section 8.10.2 :

① For interior (symmetrical) T Beams, take b the smallest of : (L = span)

   1. $b = \frac{1}{4} L$
   2. $b = b_w + 16 h_f$
   3. $b =$ center-to-center spacing of webs.

② For exterior (L-shaped) T Beams, take b the smallest of :

   1. $b = b_w + \frac{1}{12} L$
   2. $b = b_w + 6 h_f$
   3. $b = b_w + \frac{1}{2}$ (clear distance to next web).

③ For isolated T Beams, take

$$b \leq 4 b_w \qquad \text{and} \qquad h_f \geq \frac{1}{2} b_w .$$

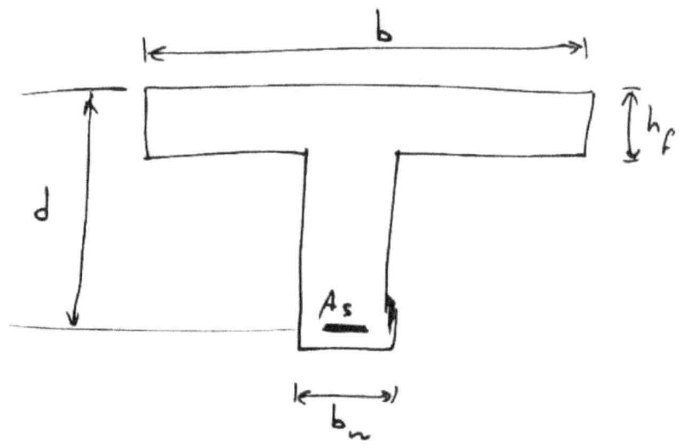

Chapter 6
Development of Reinforcement
Bond, Anchorage, and Reinforcing Details

PART I

Based on the New 1995 ACI Code

6.1 Scope of Chapter:

* **Bond stresses** are stresses between the reinforcement and the concrete.

* Bond stresses, if not limited, may produce crushing or splitting of the concrete surrounding the reinforcement, particularly if the bars are closely spaced or located near the surface of the concrete, i.e. cover is small.

* Once the reinforcement is unbounded, the beam, which behaves as if it were unreinforced, is subject to instant failure as soon as the concrete cracks.

* We will study.

(1) mechanics of bond strength

(2) development of design equations that ensure the reinforcement is solidly <u>anchored</u> to the concrete.

(3) Topics related to the termination and the detailing of reinforcement:
  * Using standard <u>hooks</u> to improve anchorage.
  * Checking local bond stresses (at simple supports and points of inflection) when the standard anchorage provisions are not fully applicable.
  * Terminating a portion of the flexural reinforcement in regions where the area of reinforcement exceeds that required for moment.
  * Detailing reinforcement to produce tough ductile members.
  * Splicing reinforcement.

6.2 Bond Stresses :

* In the design of beams, it is assumed that <u>no slippage</u> occurs between the reinforcement and the concrete.

* In other words, it is assumed that the steel will undergo the same tensile and compressive deformations as the concrete to which it is bonded.

* For the steel to deform in conjunction with the concrete, the concrete must exert shear-type stresses, termed <u>bond stresses</u>, on the surface of the reinforcement.

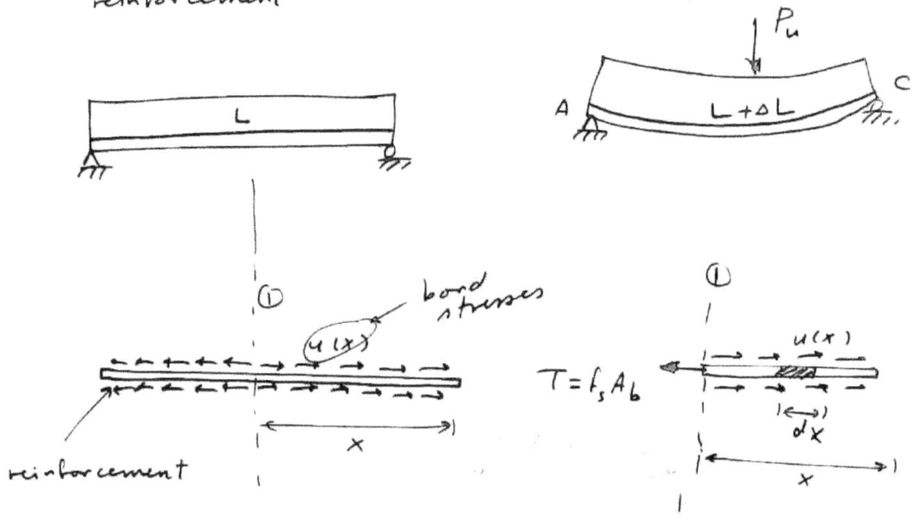

* Radial bond stresses are <u>not</u> shown in the figure above.

* If the bond stresses do not develop, the steel would remain unstressed and would not contribute to the flexural strength of the section.

* A bending failure would then occur in a sudden brittle manner when the tensile stresses in the concrete reached the modulus of rupture, the concrete behaving as if no steel were present.

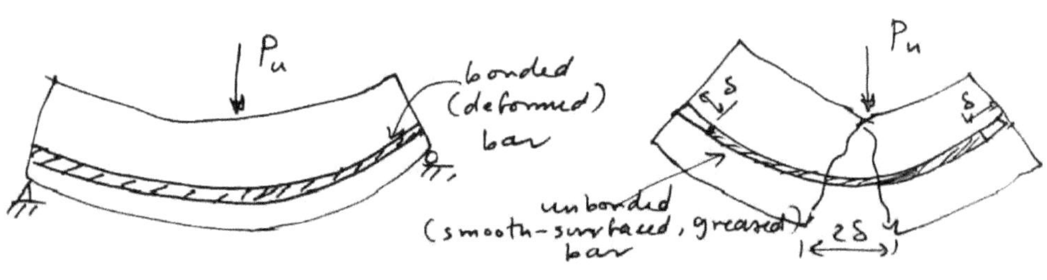

* Bond stresses vary in magnitude along the length of a reinforcing bar.

* In regions where the bending moment changes rapidly and the shear is consequently high, the tensile force in the reinforcement necessarily undergoes significant change in a short distance Δx, therefore, large bond stresses, called _flexural bond stresses_, are required to equilibrate the difference in tension ΔT between two sections.

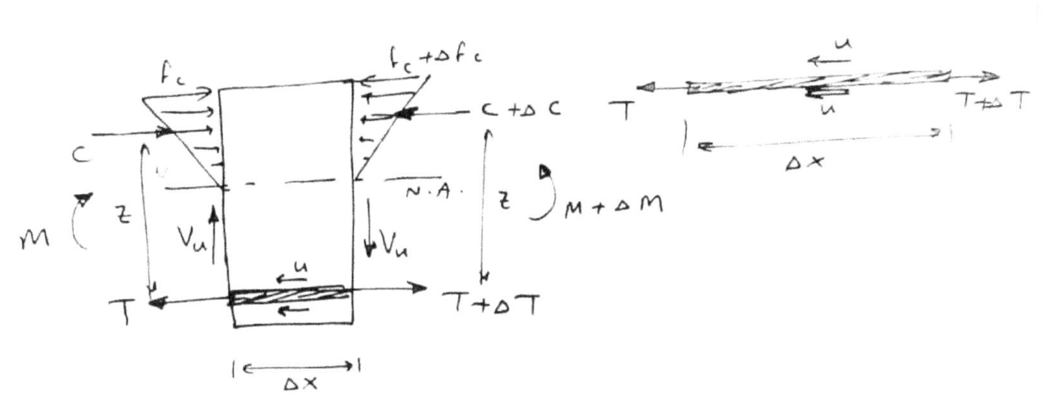

* If a crack was located at the point at which bond stresses were calculated, the actual bond stresses could deviate significantly from the computed bond stress.

* The actual _length of anchorage_, the distance between the point of maximum stress and the near end of the bar, is compared with the minimum length required for assured anchorage.

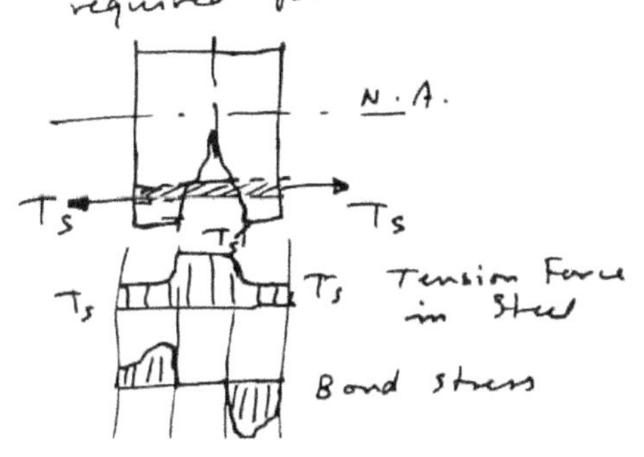

6.4 Mechanics of Bond Strength:

* Factors that contribute to bond strength are:

  1. chemical adhesion.
  2. friction.
  3. bearing of the bar deformations against the concrete.

* The contributions of each of these components of bond resistance varies with the level of stress in the reinforcement.

* When members are slightly stressed, bond resistance is due primarily to chemical adhesion. However, the bond resistance that can develop from this source is limited (1.38 MPa $\rightarrow 2.07$ MPa), and it is not a significant component of bond strength.

* After the adhesion is broken and some slight movement between the reinforcement and the concrete occurs, bond strength is supplied by both friction and by the bar deformations bearing against the concrete.

* Of these two sources of resistance to further slipping, the bearing stresses are the more significant.

* If reinforcing bars are epoxy-coated (to provide protection against corrosion), the bond stresses associated with adhesion and friction reduce sharply. For this condition, bond resistance is supplied primarily by the bearing stresses between the bar deformations and the concrete.

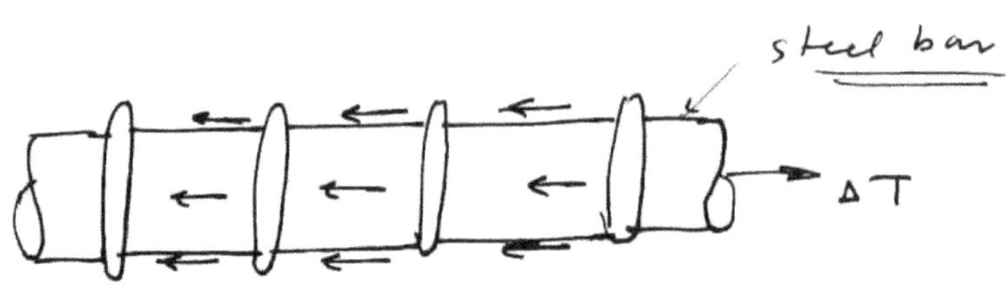

Bond Stresses Due to Friction and Chemical Adhesion

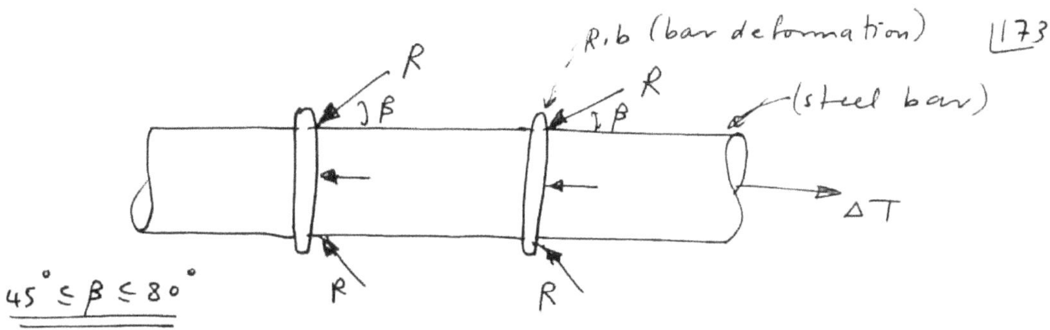

Reinforcing Bar Showing Reaction R of
Concrete on Ribs (Bar Deformations).

* the component of the bearing stresses normal to the
longitudinal axis of the bar is equal to or larger
than the longitudinal component, since $45° \leq \beta \leq 80°$.

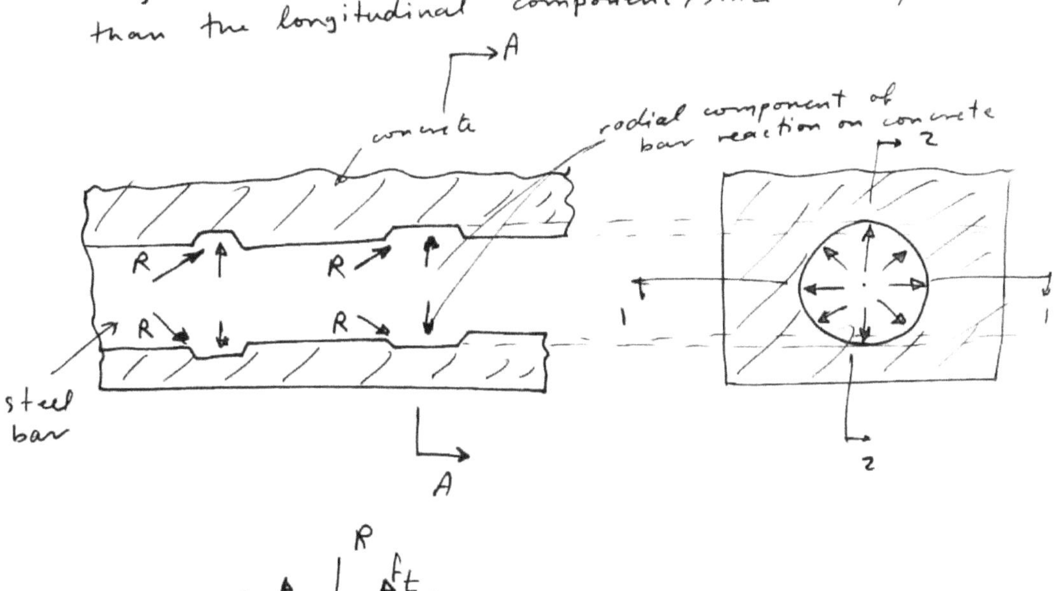

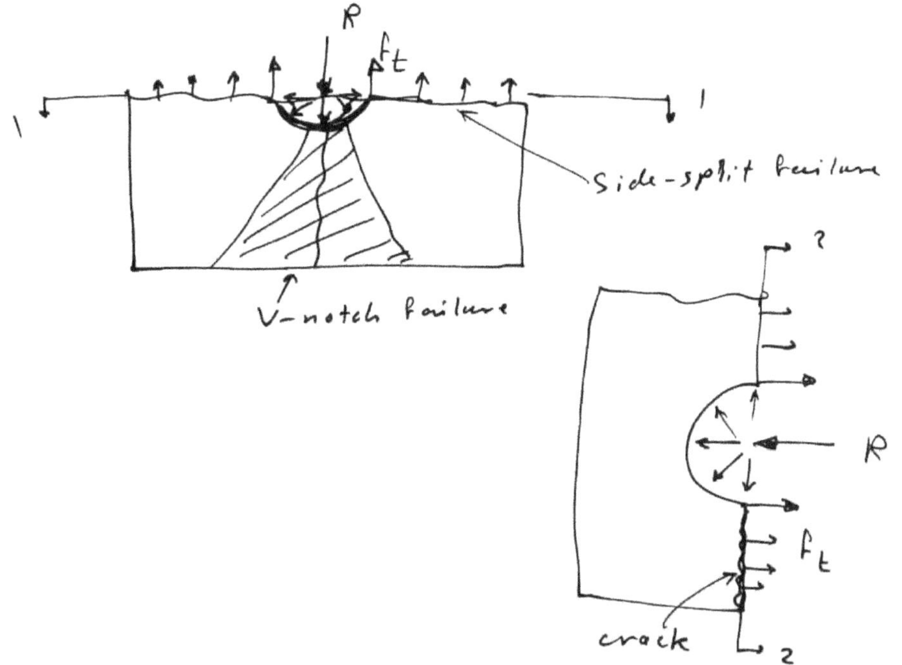

* We will discuss now the various failure modes of the cover.

① <u>horizontal side-split cracking:</u>

As the number of bars in the row increases, or equivalently, as the edge distance and spacing between bars decreases, less area of concrete is available to develop resisting tensile stresses, and the likelihood of the side-split failure increases.

② <u>V-notch failure</u>: (spalling of the cover).

this failure mode occurs if the resultant force R pushes off a triangular wedge of concrete directly below the bar.

this type of failure is most likely to occur when the bottom cover is small.

often the V-notch failure is preceded by the formation of a vertical crack extending downward from the bottom of the reinforcing bar to the outside surface of the beam.

---

* the longitudinal cracking that leads to a bond failure develops in stages. Initially, cracks of limited extent open where the local bond stresses are highest adjacent to diagonal tension or flexural cracks.

* With an increase in load, the initial cracks lengthen until they join together to form a continuous crack that extends to the end of the beam.

* Once the reinforcement crack forms, the bond between the reinforcement and the concrete is destroyed.

* As the reinforcement slips and the cover is pushed off, total collaps of the beam takes place.

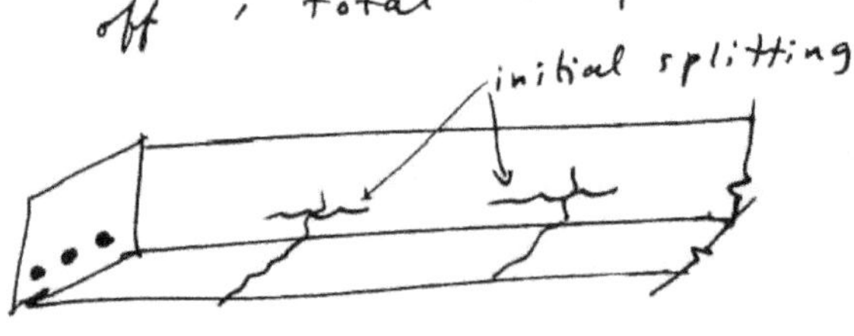

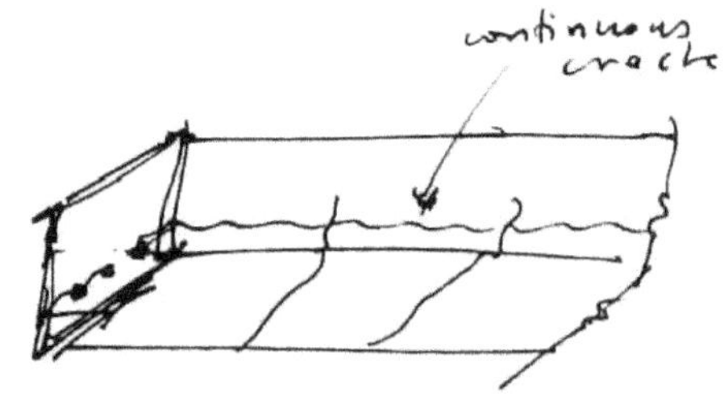

* If both the cover over the bars and the spacing of the bars are large, bond failures occur by pullout of the bar rather than by splitting of the concrete.

* Pullout failures develop when the edge cover is greater than 2.5 bar diameters and the clear spacing between bars exceeds 5 bar diameters.

* For a bar to pull out, the concrete keys between ribs must shear off, or the concrete in front of the ribs must crush.

* The likelihood of a pullout failure is increased if the concrete is weak or porous.

* The addition of transverse reinforcement along the anchorage length of a bar will produce a moderate increase in bond strength by reducing the tendency of the concrete to slip.

* The area of the transverse reinforcement — supplied by stirrups or ties — is denoted by A_{tr}.

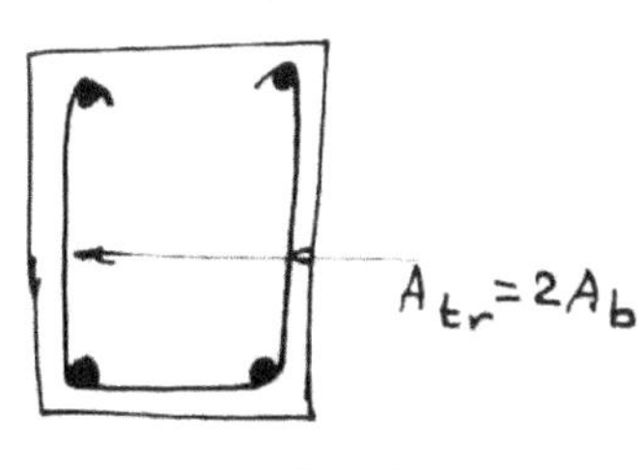

2-legged stirrups

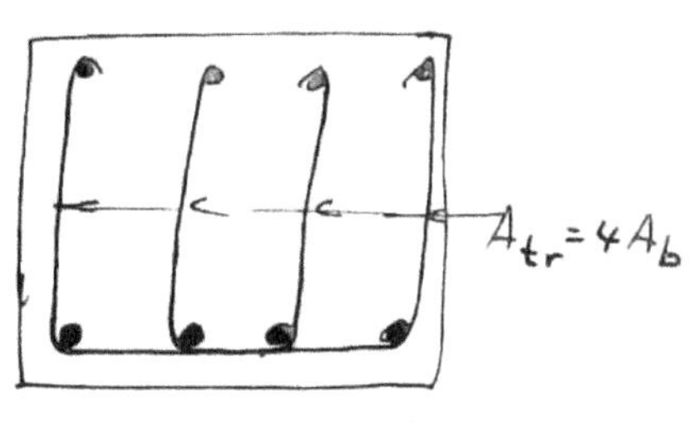

4-legged stirrups

6.5 Nominal Bond Strength:

* The nominal bond strength of a reinforcing bar can be established experimentally by measuring the force needed to produce excessive slippage or pullout of a bar embedded in concrete.

* Bond strength is influenced by several factors:
  1- flexural cracks.
  2- thickness of concrete cover.

3 - shear

4 - proximity of other bars

5 - other variables.

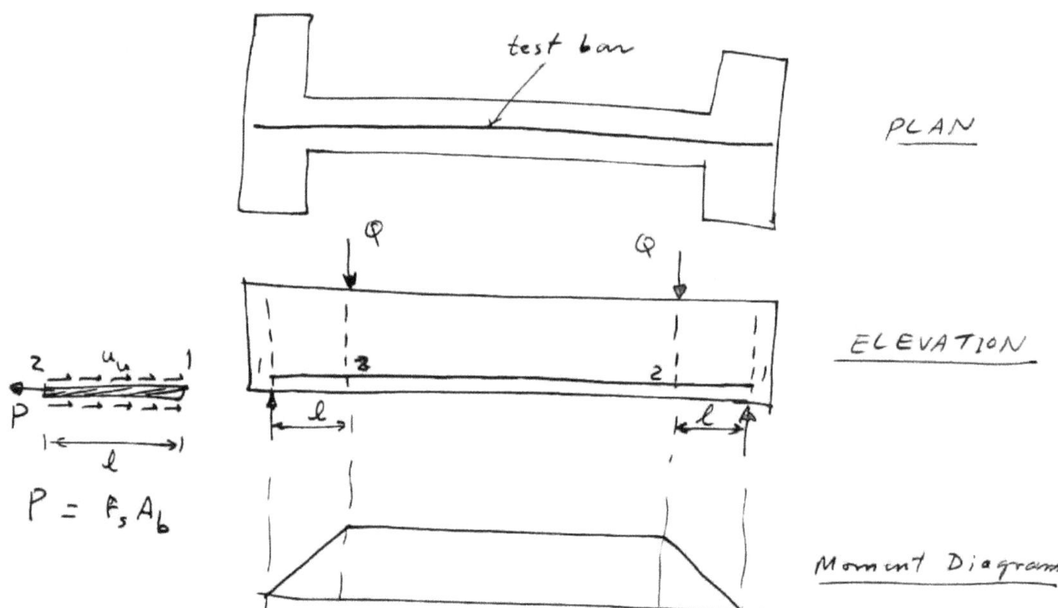

* the above setup is used experimentally to measure the nominal bond strength.

* Assume shear-type bond stresses of constant magnitude u_u to act on the bars surface over the anchorage length l.

$$\pm \Sigma F_x = 0 : \qquad u_u \, \Sigma_0 \, l = P \implies u_u = \frac{P}{\Sigma_0 \, l}$$

where

u_u : nominal bond strength

P : bar force to be anchored

Σ_0 : circumference of bar

l : length of embedment.

* Based on experimental measurements of several values of P, we have the empirical equation:

$$u_u = \frac{9.5 \sqrt{f_c'}}{d_b} \leq 800 \, psi \qquad , \quad f_c' \text{ in psi}$$

where $d_b \equiv$ bar diameter.

* If the bars are positioned in the top of a form over a depth of concrete that is 300 mm or greater, or if the concrete is made of lightweight aggregates, the bond strength will be smaller than that predicted by the above equation.

* Since it is recognized that bond failures are produced by tensile cracking, bond strength is correlated with $\sqrt{f_c'}$ rather than f_c'.

Note: Since the above equation does <u>not</u> account for the influence of the depth of cover, for the spacing between longitudinal reinforcement, or for the presence of transverse reinforcement (variables that are known to influence bond strength), experimental values of bond strength may deviate significantly from predicted values.

* To account for the influence of transverse reinforcement, the depth of bottom cover, and the spacing between bars, a reevaluation by Orangon, Jirsa, and Breen (1977) of several previous experimental bond studies on 62 beams produced the following empirical equation for u_u, the <u>ultimate bond strength</u>.

$$u_u = \left(1.2 + 3\frac{c}{d_b} + 50\frac{d_b}{l_s} + \frac{A_{tr}\, f_{yt}}{500\, s\, d_b} \right) \sqrt{f_c'}$$

where l_s = length of embedment of the longitudinal steel (in.)

d_b = bar diameter (in.)

c = the smaller of:
(1) clear bottom cover of main reinforcement
(2) half of the clear spacing between main reinforcing bars (in.).

A_{tr} = area of transverse reinforcement

f_{yt} = yield strength of the transverse reinforcement (psi)

s = spacing of transverse reinforcement in the longitudinal direction (in.)

<u>Note</u>: The above formula <u>overestimates</u> the bond strength of reinforcement in beams that are constructed with lightweight concrete, epoxy-coated reinforcing bars, or horizontal bars located at the top of beams in which at least a 300-mm depth of concrete is located under the bar.

* However, the formula can be modified by empirical factors to account for these factors.

$$\text{Chapter 6}$$
$$\text{Development of Reinforcement}$$
$$\boxed{\text{Part 2}}$$

6.6 Development Length of Tension Steel:

* The bond provisions in the ACI Code are based on the length of the bar necessary for anchorage rather than the intensity of bond stress at a point.

* <u>Length of Anchorage</u> $\equiv$ the distance from the point of peak stress to the near end of the bar.

* The ACI Code requires that the <u>length of anchorage</u> be equal to or exceed the <u>development length</u>, ℓ_d.

* The <u>development length</u>, ℓ_d, is the minimum embedment length required to anchor a bar that is stressed to the yield point, f_y.

* $\ell_d \equiv$ development length $\equiv$ minimum anchorage length.

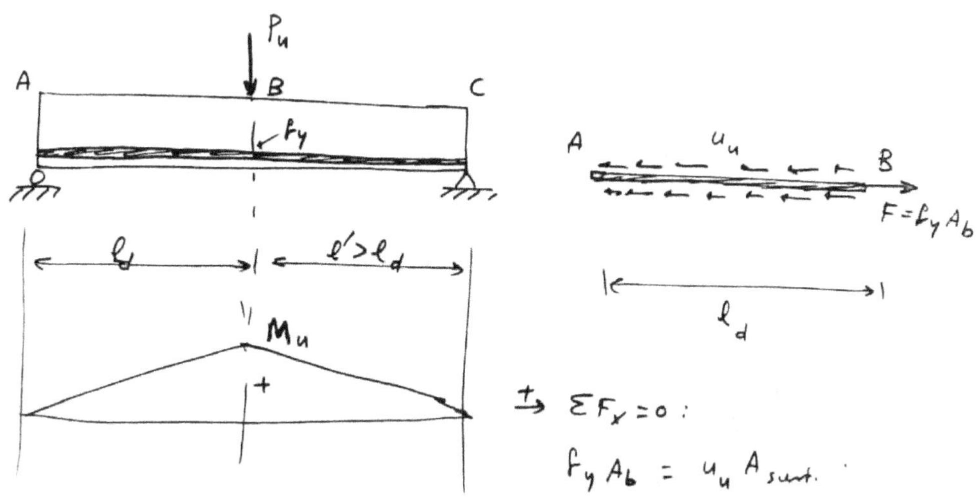

$$\xrightarrow{+} \Sigma F_x = 0 :$$
$$f_y A_b = u_u A_{surf.}$$

A_b : cross-sectional area of the bar

d_b : bar diameter.

$$A_b = \frac{\pi d_b^2}{4} \quad , \quad A_{surf.} = \pi d_b \ell_d$$

Substitute in the equilibrium equation:

$$f_y \frac{\pi d_b^2}{4} = u_u \pi d_b \ell_d$$

$$\implies \ell_d = \frac{d_b f_y}{4 u_u}$$

but $\quad u_u = \left(1.2 + 3\frac{c}{d_b} + 50\frac{d_b}{\ell_d} + \frac{A_{tr} f_{yt}}{500 s d_b}\right)\sqrt{f_c'}$

$$\text{when} \quad \ell_s = \ell_d .$$

$$\implies \boxed{\frac{\ell_d}{d_b} = \frac{f_y}{4\left[1.2 + 3c/d_b + 50 d_b/\ell_d + A_{tr} f_{yt}/500 s d_b\right)\sqrt{f_c'}}}$$

* the above equation is too complex to be used as a design equation in office practice.

* In order to produce (approximately) the same results with a simpler design equation, the 1995 ACI Code has retained most of the key variables in the above equation but adjusted the coefficients and modified certain terms.

* the <u>new equation</u> includes <u>four</u> additional basic factors that influence bond strength, $(\alpha, \beta, \gamma, \lambda)$.

* the four factors account for :

    1- position of reinforcement (i.e. top or bottom steel)

    2- epoxy coating

    3- normal or lightweight concrete

    4- bar size.

* the basic equation for development length, ℓ_d, which <u>must not be less than 300 mm</u>, is :

$$\frac{\ell_d}{d_b} = \frac{15 f_y \, \alpha \beta \gamma \lambda}{16 \sqrt{f_c'}\left[\frac{c + K_{tr}}{d_b}\right]} \qquad , \quad \begin{array}{l} f_y \text{ in } MPa . \\ f_c' \text{ in } MPa . \\ d, c \text{ in } mm . \end{array}$$

s = the maximum spacing of transverse reinforcement within the distance l_d, in (mm)

n = number of bars (or wires) being developed along the plane of splitting

c = smaller of the distance from the center of the bar to the nearest concrete surface or one-half the center-to-center spacing of the bars being developed, in (mm)

f_c' = 28-day uniaxial compressive strength. The value of $\sqrt{f_c'}$ in all equations for development length is not to exceed 100 psi (25/3 MPa) because test data on concretes with f_c' greater than 10,000 psi (69 MPa) is not available

Notes:

(1) When favorable conditions (for example, large depth of cover, wide spacing of bars, or heavy shear reinforcement) exist for bond strength, additional reduction factors can be applied to l_d (see conditions (*e*) through (*g*) in Table 6.1).

(2) Since the anchorage check is made after the cross section has been sized and both the flexural and shear reinforcement designed, d_b as well as all other properties of the cross section are known. Therefore l_d is the only unknown in Eq. (6.6).

TABLE 6.1   Modifiers for l_d/d_b in Eqs. (6.6), (6.8), (6.9), (6.10). and (6.11)†

Factor			Condition		Modifier
(*a*)	α	Bar location:	(*a*)	Top reinforcement [bars with a 12-in (300-mm) depth or more of concrete under bar]	1.3
			(*b*)	Bottom bars	1.0
(*b*)	β	Coating factor:	(*a*)	Epoxy-coated bars with cover less than $3d_b$ or clear spacing less than $6d_b$	1.5
			(*b*)	All other epoxy-coated bars	1.2
			(*c*)	Uncoated reinforcement	1.0
				Note: the product of $\alpha\beta$ need not be taken greater than 1.7	
(*c*)	γ	Bar-size factor:	(*a*)	no. 6 and smaller bars and deformed wire	0.8
			(*b*)	no. 7 and larger bars	1.0
(*d*)	λ	Lightweight-aggregate factor:	(*a*)	Lightweight-aggregate concrete	1.3
				Note: If f_{ct} is specified, λ shall be permitted to be taken as $6.7\sqrt{f_c'}/f_{ct}$ ($\sqrt{f_c'}/1.8f_{ct}$) but not less than 1.0	
			(*b*)	Normal weight concrete	1.0
(*e*)	Reduction for excess reinforcement:				$A_{s,req}/A_{s,sup}$
(*f*)	Reduction for large cover and wide bar spacing:				0.8
	For no. 11 (no. 35) and smaller bars with a *clear* spacing of not less than $5d_b$ and with cover not less than $2.5d_b$ from the face of member to edge of bar (measured in the plane of the bars)				
(*g*)	Heavily confined reinforcement:				0.75
	For reinforcement enclosed (1) within a spiral of diameter $\frac{1}{4}$ in (6 mm) or more and with a pitch of not more than 4 in (100 mm), or (2) within no. 4 (no. 15) or larger circular ties spaced not more than 4 in (100 mm) on center, or (3) within no. 4 (no. 15) ties or stirrups spaced not more than 4 in (100 mm) on center and arranged so that alternate bars are restrained by the corner of a tie or hoop with an included angle of not more than 135°.				

† In Table 6.1 the use of the reduction factors of 0.8 and 0.75 listed for conditions (*f*) and (*g*), respectively, is optional and left to the discretion of the designer.

TABLE 6.2 Equations for required values of l_d/d_b

Conditions: cover, bar spacing, stirrups	Bar Size	
	No. 6 (no. 20) and smaller bars and deformed wire	No. 7 (no. 25) and larger bars
Case A (1) Clear spacing of bars being developed not less than d_b, and clear cover not less than d_b, or values given in Table 3.4, and stirrups or ties not less than Code minimum (see Table 4.1); or (2) clear cover not less than d_b, and clear spacing of bars being developed not less than $2d_b$.	$\dfrac{l_d}{d_b} = \dfrac{f_y \alpha \beta \lambda}{25 \sqrt{f_c'}}$ USCU (6.8) $\dfrac{l_d}{d_b} = \dfrac{f_y \alpha \beta \lambda}{2 \sqrt{f_c'}}$ SI (6.8a)	$\dfrac{l_d}{d_b} = \dfrac{f_y \alpha \beta \lambda}{20 \sqrt{f_c'}}$ USCU (6.9) $\dfrac{l_d}{d_b} = \dfrac{5 f_y \alpha \beta \lambda}{8 \sqrt{f_c'}}$ SI (6.9a)
Case B Smaller cover, closer spacing of bars, or other less favorable conditions	$\dfrac{l_d}{d_b} = \dfrac{3 f_y \alpha \beta \lambda}{50 \sqrt{f_c'}}$ USCU (6.10) $\dfrac{l_d}{d_b} = \dfrac{3 f_y \alpha \beta \lambda}{4 \sqrt{f_c'}}$ SI (6.10a)	$\dfrac{l_d}{d_b} = \dfrac{3 f_y \alpha \beta \lambda}{40 \sqrt{f_c'}}$ USCU (6.11) $\dfrac{l_d}{d_b} = \dfrac{15 f_y \alpha \beta \lambda}{16 \sqrt{f_c'}}$ SI (6.11a)

A slightly simpler approach for establishing the development length l_d for several common design conditions is also permitted by the ACI Code. In this alternate procedure, the Code recognizes that many practical designs utilize values of cover and spacing between bars as well as stirrup spacing that results in values of $[(c + K_{tr})/d_b]$ equal to or greater than 1.5. For example, cross sections in which the clear cover is not less than d_b and the clear spacing between bars being developed is not less than $2d_b$ would satisfy this requirement. A second condition that would produce a value of $[(c + K_{tr})/d_b]$ equal to or greater than 1.5 would exist if (1) the clear spacing between bars is not less than d_b, (2) the spacing of stirrups along the entire development length is not more than the distances specified in Table 4.1, and (3) the minimum cover requirements as listed in Table 3.5 are satisfied.

To cover these cases, as well as the influence of bar size on development length, we can use the *Case A* equations [see Eqs. (6.8) and (6.9)] in the first line of Table 6.2. For less favorable conditions, *Case B* equations [see Eqs. (6.10) and (6.11)] of Table 6.2 control. These latter equations, which require a 50 percent increase in development length, are established by multiplying the equations in the first line by a factor of 1.5. The equivalent equations in metric units are given as Eqs. (6.8a), (6.9a), (6.10a), and (6.11a).

Development lengths for the vast majority of beams containing stirrups and designed with the minimum cover over the flexural steel specified by the ACI Code will be controlled by the *Case A* equations. Since the steel and concrete strengths used throughout most sections of a structure are constant, we can establish, for this common situation, simple equations for development length expressed as a *constant* times the bar diameter d_b (see Example 6.5).

If bars in a given row terminate at different points, the anchorage regions (where the bond stresses are greatest) for each set of similar-length bars are in different locations. If the designer alternates the long and short bars in a particular row, the effective spacing used to select the appropriate modifiers in Table 6.1 can be taken as the distance between bars of the same length.

but $\qquad \dfrac{c + K_{tr}}{d_b} \leq 2.5$

[in order to ensure that pullout failure will not occur].

$K_{tr} \equiv$ transverse reinforcement index

$$K_{tr} = \frac{A_{tr}\, f_{yt}}{260\, s n}$$

where

A_{tr} : total cross-sectional area of all transverse reinforcement which is within the stirrup spacing s , (in mm^2).

f_{yt} : yield strength of transverse reinforcement (in MPa)

s : maximum spacing of transverse reinforcement within the distance l_d (in mm).

n : number of bars being developed along the plane of splitting.

c : smaller of the distance from the center of the bar to the nearest concrete surface or one-half the center-to-center spacing of the bars being developed (in mm).

f_c' : 28-day uniaxial compressive strength.

Note : the value of $\sqrt{f_c'}$ in all equations for development length is <u>not</u> to exceed $\dfrac{25}{3}$ MPa because test data on concretes with $f_c' > 69$ MPa is not available.

---

* the four factors $\alpha, \beta, \gamma, \lambda$ are <u>modifiers</u> for l_d/d_b and are obtained from the following table: (<u>Table 6.1</u>)

* A slightly simpler approach for establishing the development length l_d for several common design conditions is also permitted by the ACI Code.

(See Table 6.2)

* If the designer alternates the long and short bars in a particular row, the effective spacing used to select the appropriate modifiers in Table 6.1 can be taken as the distance between bars of the same length.

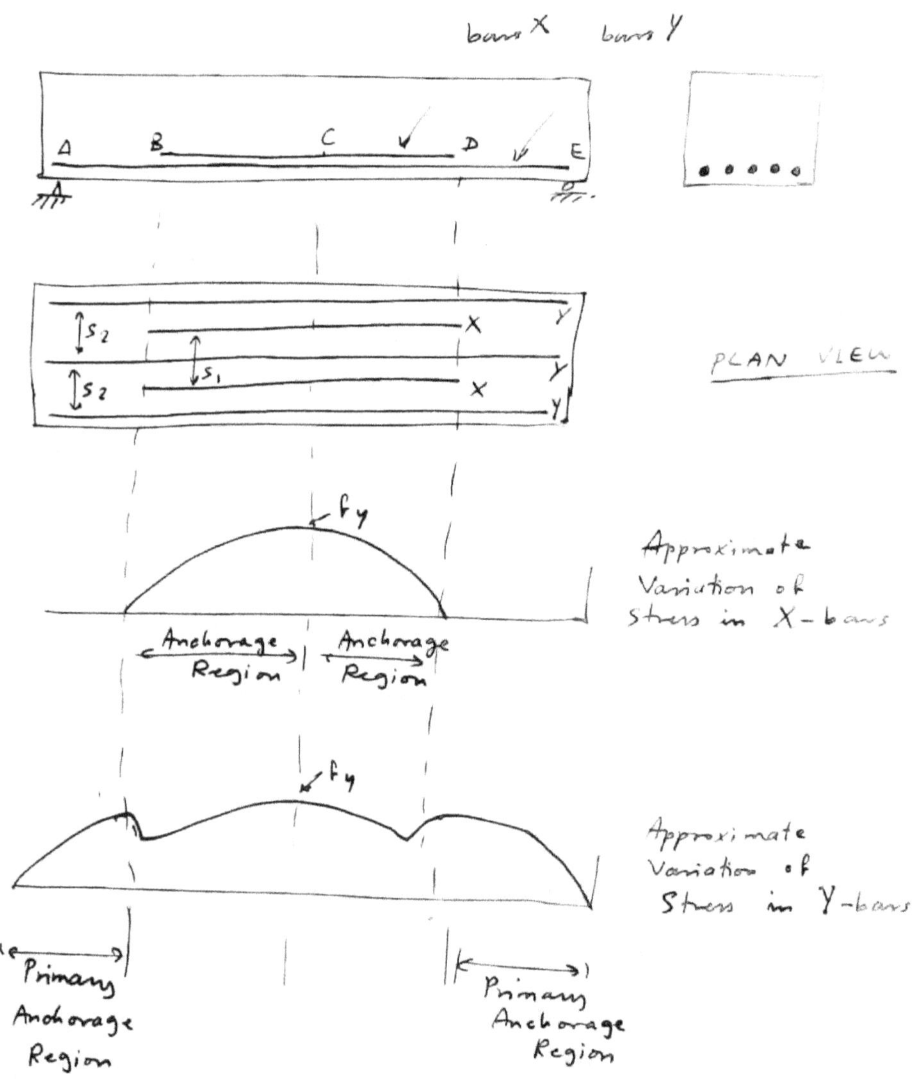

Notes on Bundled Bars

* If bars are bundled, the ACI Code (Section 12.4.1) specifies that although the development length is still based on the area of an individual bar, it is to be increased 20% for a three-bar bundle and 33% for a four-bar bundle; no increase is required for a two-bar bundle.

* The ACI Code states that when the designer determines the appropriate modifiers in Table 6.1, a unit of bundled bars is to be treated as a single bar with an effective diameter that supplies the same area as that of the bars in the bundle.

* For example, a four-bar bundle of ∅28 mm, which contains $A_s = 2463$ mm² of steel, would be approximated by a single bar with a diameter of 56 mm.

Summary of Procedure to Compute the Required Development Length, l_d:

① Draw a sketch of the cross-section and establish the clear distance between bars as well as the clear cover.

② Determine the required development length l_d using the equation or the appropriate expression in Table 6.2. The length l_d must not be less than 300 mm.

③ ACI Code requirements for anchorage are satisfied if the available anchorage length equals or exceeds l_d.
If l_d exceeds the available anchorage length, the designer can add a hook to the end of bars, weld the end of the bar to a plate or structural steel shape, or add additional stirrups or a spiral

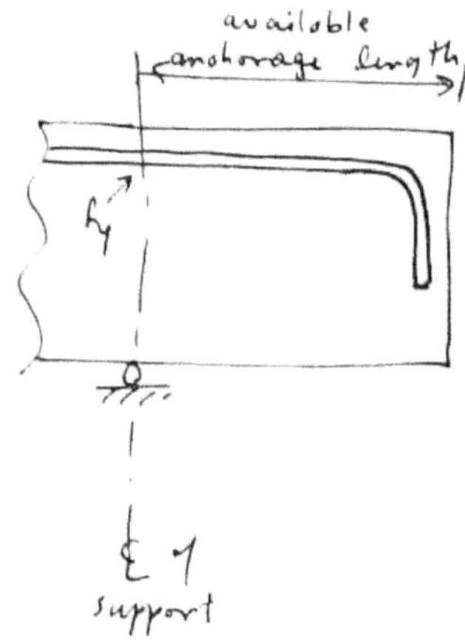

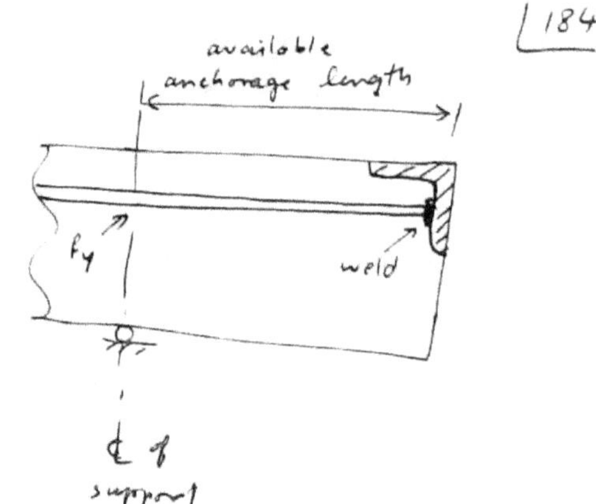

Example 1 :

A beam constructed from lightweight aggregate concrete
is reinforced with three $\phi25\,mm$ bars in a single row.
The ends of the longitudinal steel terminate $50\,mm$ from
the ends of the beam.

(a) Determine if the reinforcement satisfies ACI Code
requirements for development length.

(b) If the Code requirements are _not_ satisfied, would
increasing the width of the beam to $300\,mm$
produce an acceptable design ?

Assume the area of steel supplied equals the area
of steel required. Take $f_y = 415\,MPa$ and $f_c' = 20\,MPa$.

Solution :

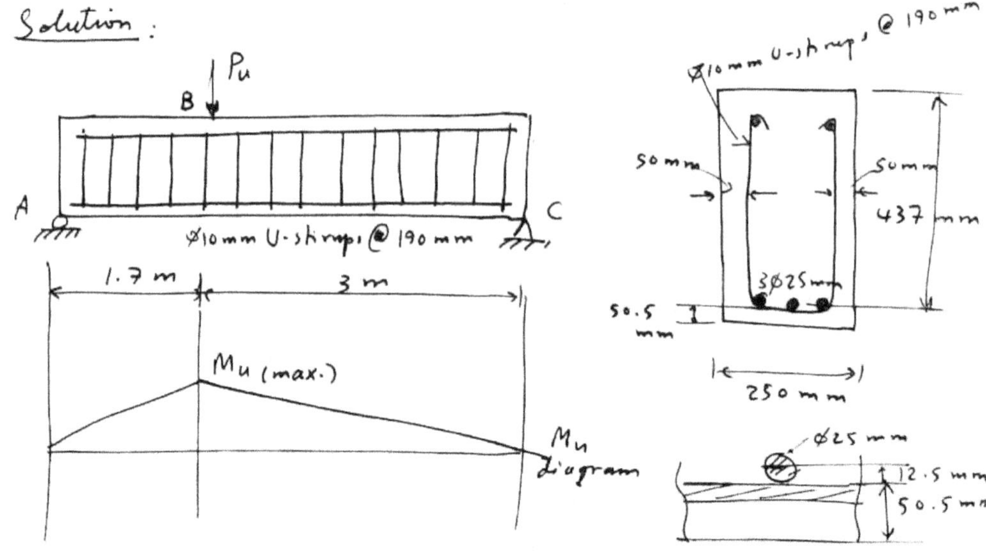

(a) The bottom reinforcement is stressed to its yield $\boxed{125}$
point, f_y directly under the concentrated load at B.

Available Anchorage length $= 1.7 - \dfrac{50}{1000} = 1.65\ m = 1650\ mm$

Compute the basic development length ℓ_d

$$\frac{\ell_d}{d_b} = \frac{15 f_y\ \alpha\beta\delta\lambda}{16\sqrt{F_c'}\ \left[\dfrac{c + K_{tr}}{d_b}\right]}$$

Find the four modifiers $\alpha, \beta, \delta, \lambda$ from Table 6.1

$\quad \alpha = 1 \qquad$ (bottom bars)

$\quad \beta = 1 \qquad$ (uncoated reinforcement)

$\quad \delta = 1 \qquad \left[\ \phi 25\ mm\ \ \overset{(\#8)}{are\ no.8}\ bars\ > no.7\ \right]$

$\quad \lambda = 1.3 \qquad$ lightweight concrete.

<u>Evaluate c</u>: Use the smaller of:

$\quad$ (1) distance from center of bar to outside surface
$\quad$ (2) one-half the center-to-center distance
$\qquad\quad$ between bars

(1) $\quad c = 50.5 + \dfrac{25}{2} = 63\ mm$.

(2) $\quad$ center-to-center distance $= \dfrac{250 - 2(50) - 25}{2} = 62.5$

$\qquad\qquad c = \dfrac{62.5}{2} = 31.25\ mm$

$\Longrightarrow \quad$ Take $\quad c = 31.25\ mm$.

Evaluate $\quad K_{tr} = \dfrac{A_{tr} f_{yt}}{260\ sn} = \dfrac{2\left(\dfrac{\pi(10)^2}{4}\right) \cdot 415}{(260)(190)(3)} = 0.44$

Evaluate $\left(\dfrac{c + K_{tr}}{d_b}\right) = \dfrac{31.25 + 0.44}{25} = 1.27 \; < 2.5$

$\qquad\qquad\qquad\qquad\qquad\qquad \underline{\underline{o.k.}}$

$$\therefore \quad \frac{l_d}{d_b} = \frac{15 f_y \, \alpha \beta \gamma \lambda}{16 \sqrt{f_c'} \left[\frac{c + k_{tr}}{d_b} \right]} = \frac{(15)(415)(1)(1)(1)(1.3)}{16 \sqrt{20} \quad (1.27)}$$

$$\Rightarrow \quad \frac{l_d}{d_b} = 89.05$$

But $d_b = 25$ mm $\Rightarrow l_d = (89.05)(25) = 2226$ mm

$\therefore \quad l_d >$ available anchorage length $= 1650$ mm.

this is unacceptable.

Try to use the more conservative expression from Table 6.2:

$$\frac{l_d}{d_b} = \frac{5 f_y \, \alpha \beta \lambda}{8 \sqrt{f_c'}} = \frac{5(415)(1)(1)(1.3)}{8 \sqrt{20}} = 75.4$$

$$\Rightarrow \quad l_d = (25)(75.4) = 1885 \text{ mm} > 1650.$$

$$\Rightarrow \quad \text{this is still unacceptable.}$$

(b) Increase the width of the beam to 300 mm.

center-to-center spacing between bars $= \dfrac{300 - 2(50) - 25}{2} = 87.5$ mm

$$c = \frac{87.5}{2} = 43.75 \text{ mm} \quad \text{(still smaller)}.$$

$$\frac{c + k_{tr}}{d_b} = \frac{43.75 + 0.44}{25} = 1.76 < 2.5 \quad \text{o.k.}$$

$\therefore$

$$\frac{l_d}{d_b} = \frac{15 f_y \, \alpha \beta \gamma \lambda}{16 \sqrt{f_c'} \left[\frac{c + k_{tr}}{d_b} \right]} = \frac{(15)(415)(1)(1)(1)(1.3)}{16 \sqrt{20} \quad (1.76)}$$

$$\Rightarrow \quad \frac{l_d}{d_b} = 64.3$$

$$\Rightarrow \quad l_d = (25)(64.3) = 1606 \text{ mm} < 1650 \text{ mm}. \quad \text{o.k.}$$

$$\Rightarrow \quad \text{anchorage is o.k.}$$

Example 2 :

An area of top flexural steel equal to $1300\ mm^2$ is required to carry the negative moment in the beam at support B shown in the figure. If two no.30 bars ($A_s = 1400\ mm^2$) are used as reinforcement, is sufficient embedment length available in the cantilever to satisfy ACI Code requirements for anchorage length? The cantilever is reinforced with $\phi10\ mm$ stirrups spaced $200\ mm$ on center throughout the entire length of the beam. Two no.15 bars in the bottom of the compression zone are added to reduce creep and to anchor and position the stirrups. The main flexural reinforcement has $50\ mm$ of clear cover on all sides. Take $f_c' = 28\ MPa$ and $f_y = 400\ MPa$. $\beta = \lambda = 1$.

Solution :

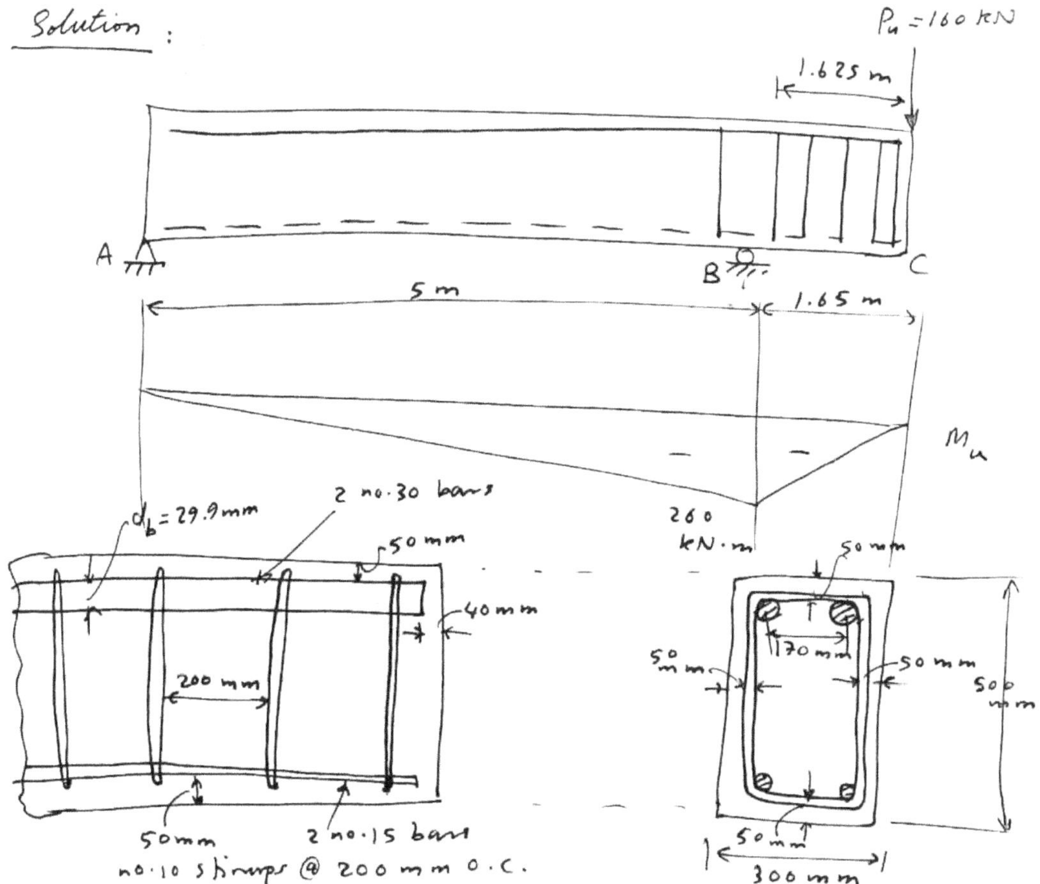

Available anchorage length $= 1650 - 40 = 1610$ mm.

The critical region for anchorage is between B and C.

* Clear cover $= 50$ mm $> d_b = 29.9$ mm

* Clear spacing $= 300 - 2(50) - 2(30) = 140$ mm $> 2 d_b \approx 60$ mm .
  (between bars)

Use the appropriate equation from Table 6.2 .

Case A (2): , ($\alpha = 1.3$ for top bars)

$$\frac{l_d}{d_b} = \frac{5 f_y \alpha \beta \lambda}{8 \sqrt{f_c'}} = \frac{5 \times (400)(1.3)(1)(1)}{8 \sqrt{28}} = 61.42$$

$$\Rightarrow \quad l_d = 61.42 d_b = 61.42(29.9) = 1836 \text{ mm}$$

However, since $A_{s(used)} > A_{s(required)}$,

we need to modify l_d for excess steel area.

See Table 6.1 (item (e)):

$$\text{reduction} = \frac{A_s (required)}{A_s (supplied)} = \frac{1300}{1400}$$

$$\therefore \quad l_d = 1836 \left(\frac{1300}{1400} \right) = \underline{\underline{1705 \text{ mm}}}$$

$\therefore$ Available anchorage length $= 1610$ mm $< l_d = 1705$ mm.

Section does __not__ satisfy ACI Code anchorage requirements.

Note :
If the designer increases the stirrups to $\phi 16$ mm and decreases their spacing to 100 mm, he or she can use the 0.75 reduction factor, permitted by condition (g) in Table 6.1 , to reduce the magnitude of l_d .

New $l_d = 1705 \times 0.75 = 1279$ mm .

$\therefore$ This solution satisfies ACI Code requirements for anchorage .

Alternatively : Use hooks.

Example 3 :

The beam shown in the figure is reinforced for positive moment with four epoxy-coated ⌀25 mm bars. The bars terminate at the center of the supports. Two concentrated loads of the same magnitude, positioned at the beam's third points, represent the main loads, i.e. the dead weight of the beam has been neglected in constructing the moment diagram. If the bars are formed into two 2-bar bundles, is their embedment length adequate to satisfy ACI Code anchorage requirements? the area of steel supplied is equal to the area of steel required, and there are 50 mm of clear cover on all sides of the longitudinal steel. Normal weight concrete ($\lambda=1$), $f_c' = 20$ mPa, and $f_y = 415$ mPa.

Solution :

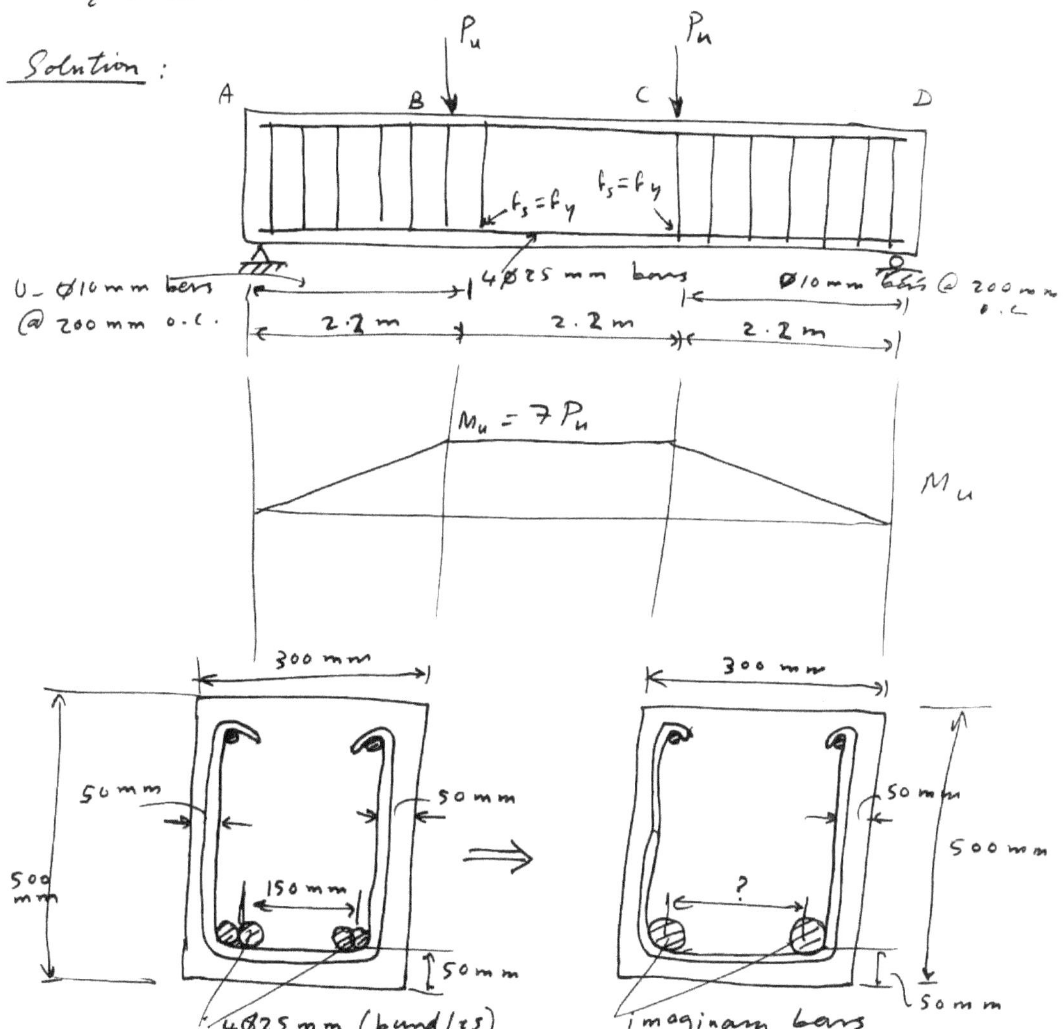

Available anchorage length = 2.2 m $\equiv 2200$ mm.

(between A and B, or C and D, by __symmetry__)

ACI Code/Section $12.4.1$ specifies that we treat each two-bar bundle as an __imaginary single bar__ with an __effective diameter__ d_{eff} that supplies the same steel area.

$$\frac{\pi d_{eff}^2}{4} = 2\left(\frac{\pi (25)^2}{4}\right) \Rightarrow d_{eff} = 35.36 \text{ mm}.$$

Since $d_{eff.} = 35.36 >$ diameter of no. 7 bar

$$\Rightarrow \gamma = 1.$$

Cover and spacing satisfy Table 6.2 (Case A (1)):

$$\alpha = 1 \quad (\text{bottom bars}).$$

For α: (epoxy-coated bars):

Distance from center of imaginary bar to outside surface:

$$= 50 + \frac{35.36}{2} = 67.68 \text{ mm} < 3d_b = 3(35.36) = 106.08 \text{ mm}.$$

Distance center-to-center of imaginary bars

$$= 300 - 2(50) - 35.36 = 164.64 \text{ mm} < 6d_b = 6(35.36) = 212.16.$$

$$\Rightarrow \beta = 1.5$$

$$\therefore \frac{l_d}{d_b} = \frac{5 f_y \alpha \beta \lambda}{8\sqrt{f_c'}} = \frac{5(415)(1)(1.5)(1)}{8\sqrt{20}} = 86.997$$

$$\Rightarrow l_d = 86.997 d_b = 86.997(25) = 2175 \text{ mm}. \quad (\phi 25 \text{ mm})$$

Available anchorage length $= 2200$ mm $> l_d = 2175$ mm.

$$\Rightarrow \text{Anchorage is o.k.}$$

Notes :

(1) If we had used either a three-bar or a four-bar bundle, ACI Code/Section 12.4.1 requires that we increase the preceding value of l_d by 20% or 33%, respectively. For either of these cases, the bars would **not** satisfy ACI Code requirements for anchorage.

(2) For computation of the crack control parameter z (see Chapter 4), a bundle of bars must be treated as a single bar.

Question: Calculate $z = ?$ for Example 3

Answer: (*) it satisfies ACI Code crack control requirements for interior exposure, **but not** for exterior exposure.

Example 4 :

Determine if the transverse reinforcement ($\phi 16$ mm @ 200 mm) in the one-way wall footing shown in the figure satisfies the ACI Code requirements for anchorage. The bars, which are epoxy-coated, extend to within 75 mm of the edge of the footing. The peak stress is assumed to occur on a section at the face of the wall. Take $f_c' = 20$ MPa, $f_y = 415$ MPa and $A_{s\,(supplied)} = A_{s\,(required)}$.

Solution :

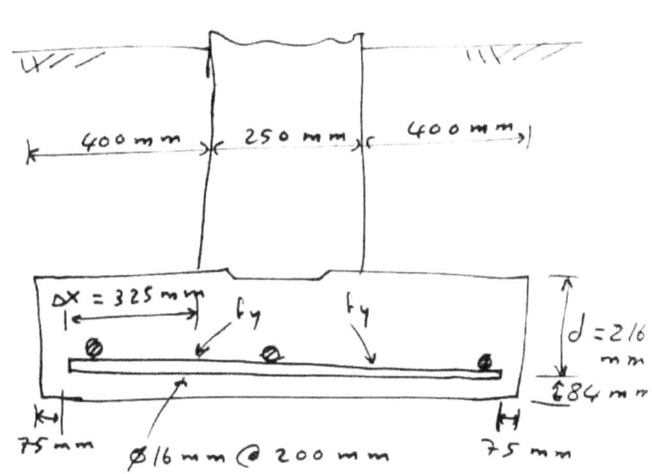

$$\frac{\ell_d}{d_b} = \frac{15 f_y \; \alpha\beta\gamma\lambda}{16 \sqrt{f_c'} \left[\dfrac{c + K_{tr}}{d_b}\right]}$$

(1) Evaluate (c):

    center of bar to surface of concrete = 84 mm

    center-to-center spacing between bars = 200 mm .
    ↳ half this distance $= \dfrac{200}{2} = 100$ mm .

    $\Longrightarrow$ Take $c = 84$ mm .

(2) No stirrups are used $\Longrightarrow A_{tr} = 0$

$$\Longrightarrow \quad K_{tr} = \frac{A_{tr} f_{yt}}{260\, s\, n} = 0$$

$$\Longrightarrow \quad \frac{c + K_{tr}}{d_b} = \frac{84 + 0}{16} = 5.25 \; \not< \; 2.5$$

$$\Longrightarrow \quad \boxed{Use \; 2.5} \checkmark$$

$\alpha = 1$ (bottom bars)

$\beta = 1.2$ (bar spacing = 200 mm $> 6 d_b = 6(16) = 96$ mm .)
                 (cover = 84 mm $> 3 d_b = 3(16) = 48$ mm) .

$\gamma = 0.8$ (bar size $\varnothing 16$ mm $<$ no. 6) .

$\lambda = 1$ (normal weight concrete) .

$$\Longrightarrow \quad \frac{\ell_d}{d_b} = \frac{15 f_y \; \alpha\beta\gamma\lambda}{16 \sqrt{f_c'}\left[\dfrac{c+K_{tr}}{d_b}\right]} = \frac{15(415)(1)(1.2)(0.8)(1)}{16\sqrt{20}\;(2.5)}$$

$$= 33.41$$

$$\Longrightarrow \quad \ell_d = 33.41\, d_b = 33.41(16) = 535 \text{ mm} .$$

$\ast$ Since we have large spacing and cover, we can reduce ℓ_d .
    see Table 6.1 (f) :
        Reduced $\ell_d = 0.8(535) = 428$ mm .

Available Anchorage Length = $325\,mm < l_d = 428$.

$\Rightarrow$ Anchorage is <u>not</u> satisfactory.

<u>Suggested Solutions</u>:

(1) Reduce the bar size.

    e.g. $\phi 12\,mm$ @ $120\,mm$ spacing $\Rightarrow A_s = 942\,mm^2$.

    [<u>Note</u>: original $A_s = 1005\,mm^2$].
      (should be very close).

    $c = 84^{mm}$ (still the same).

$$\frac{c + K_{tr}}{d_b} = \frac{84 + 0}{12} = 7.0 \not< 2.5 \Rightarrow \text{Use } 2.5.$$

$$\frac{l_d}{d_b} = 33.41 \Rightarrow l_d = 33.41\,(12) = 401\,mm.$$

    Reduced $l_d = 0.8\,(401) = 321\,mm$.

$\Rightarrow$ Available Anchorage Length $= 325\,mm > l_d = 321\,mm$

$\Rightarrow$ o.k.

---

(2) Widen the footing.

(3) Increase the concrete strength.

(4) Hook the ends of the bar.

<u>Example 5</u>:

The floor system of a building is to be constructed of normal weight concrete with $f_c' = 28\,MPa$ and steel with yield strength $f_y = 415\,MPa$. If the reinforcement is uncoated, develop expressions for development length in terms of the bar diameter d_b and a constant, assuming all bars satisfy the <u>Case A</u> conditions in Table 6.2.

Solution:

$$\alpha = \begin{cases} 1 & \text{for bottom bars} \\ 1.3 & \text{for top bars} \end{cases}$$

$\beta = 1$ (uncoated reinforcement).

$\lambda = 1$ (normal weight concrete).

No.6 and smaller bars (bottom bars)	No.7 and larger bars (bottom bars)
No.6 bars $\doteq$ $\emptyset$19.05 mm.	No.7 bars $\doteq$ $\emptyset$22.22 mm.
$\dfrac{l_d}{d_b} = \dfrac{f_y \, \alpha \beta \lambda}{2\sqrt{f_c'}}$	$\dfrac{l_d}{d_b} = \dfrac{5 f_y \, \alpha \beta \lambda}{8\sqrt{f_c'}}$
$= \dfrac{(415)\,(1)(1)(1)}{2\sqrt{28}}$	$= \dfrac{5(415)\,(1)(1)(1)}{8\sqrt{28}}$
$= 39.21$	$= 49.02$
$\Rightarrow \boxed{l_d = 39.21\, d_b}$ bottom bars	$\boxed{l_d = 49.02\, d_b}$ bottom bars.
For top bars, multiply by $\underline{1.3}$:	
$\Rightarrow \boxed{l_d = 50.97\, d_b}$ top bars	$\boxed{l_d = 63.73\, d_b}$ top bars.

6.7 Development Length of Compression Steel:

* Smaller development lengths are required for compression reinforcement than for tension reinforcement.

     (because of lack of cracking and bearing).

* ACI Code / Section 12.3 :

     Basic development length in compression equals:

$$\boxed{l_{db} = \dfrac{f_y \, d_b}{4\sqrt{f_c'}}}$$

f_c', f_y in MPa

but $\underline{not}$ less than $0.044\, f_y\, d_b$ or 200 mm.

d_b in mm.

<u>Note</u>: If excess steel area is supplied, l_{db} can be further reduced by the ratio $\dfrac{A_s \,(required)}{A_s \,(supplied)}$.

<u>Note</u>: If the reinforcement is encased in a <u>spiral</u> (the diameter must be at least 6 mm, and the <u>pitch</u> must not exceed 100 mm), or within no. 15 metric ties ($\approx \phi 16$ mm) spaced not more than 100 mm on center, then the ACI Code permits a 25% reduction in the development length.

<u>Note</u>: When bars are <u>bundled</u>, the development length of an individual bar is computed in the same manner as for a single bar but is then increased 20% for a three-bar bundle or 33% for a four-bar bundle. No increase is required for a two-bar bundle.

Example 6 :

Four $\phi 25$ mm dowels stressed to f_y are required to transfer the axial compression force in a pier into the footing shown in the figure. Determine the minimum extension of the dowels into the footing. Use $f_c' = 24$ MPa, $f_y = 415$ MPa, and $\phi 12$ mm ties at 200 mm on center.

Although dowels are typically bent so that they can be positioned securely by wiring to the footing reinforcement, the bend does <u>not</u> contribute any resistance to compression.

<u>Solution</u>:

$$l_{db} = \frac{f_y d_b}{4\sqrt{f_c'}} = \frac{415\,(25)}{4\,\sqrt{24}}$$

$$= 529 \text{ mm}.$$

but <u>not</u> less than:

$$0.044 f_y d_b = 0.044\,(415)(25)$$

$$= 456 \text{ mm or } 200 \text{ mm}.$$

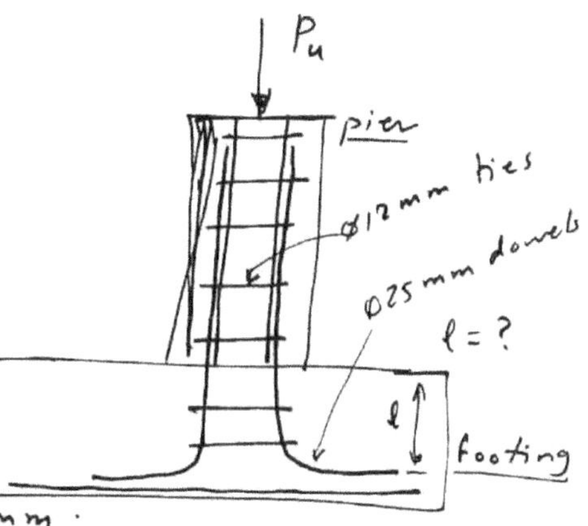

$\Rightarrow$ use $l_d = 529$ mm.

$\Rightarrow$ Depth of footing $\approx 529 + 75 = 599$ mm $\Rightarrow 600$ mm.

<u>Note</u> : If this depth is not available, you can use a larger number of smaller dowels to reduce the development length.

e.g. : Assume that two $\phi 20$ mm are substituted for each $\phi 25$ mm dowel. Determine the minimum required extension into the footing.

$$A_{s \atop (supplied)} = 2 \left[\frac{\pi (10)^2}{4} \right] = 628 \text{ mm}^2.$$

$$l_d = \frac{f_y \, d_b}{4 \sqrt{f_c'}} = \frac{(415)(20)}{4\sqrt{24}} = 424 \text{ mm}.$$

$$A_{s\,(required)} = \frac{\pi (25)^2}{4} = 491 \text{ mm}^2.$$

$$\text{Reduced } l_d = 424 \, \frac{A_{s\,(required)}}{A_{s\,(supplied)}} = 424 \left(\frac{491}{628}\right)$$
$$= 332 \text{ mm}.$$

but not less than $\quad 0.044 \, f_y \, d_b = 0.044 \,(415)(20)$
$$= 365 \text{ mm}.$$
$$\text{or} \quad 200 \text{ mm}.$$

$$\Rightarrow \quad \text{Take } l_d = \frac{365 \text{ mm}.}{332} \text{ mm}.$$

<u>Note</u> : Another solution to reduce the development length is to reduce the spacing of the $\phi 12$ mm ties from 200 mm to 100 mm.

$\Rightarrow \quad$ a **25%** reduction is permitted by ACI Code.

$\therefore \quad$ Reduced $l_d = 529 \times 0.75 = 397$ mm.

6.8 Standard Hooks:

* If a beam's dimensions are limited so that a tension bar cannot be extended in a straight line a distance equal to the required development length, a **hook** may be added to provide additional anchorage capacity.

* Either a $90°-$ or a $180°-$ bend may be used to form a hook.

* Both types of hooks have exactly the same anchorage capacity and are designed in the same manner.

* The $180°-$hook is most suited to shallow members

* The $90°-$hook is often used when horizontal reinforcement in one member is made continuous with the vertical reinforcement in a second member.

Note : Experimental studies indicate hooks are **not effective** in anchoring bars stressed in <u>compression</u>.

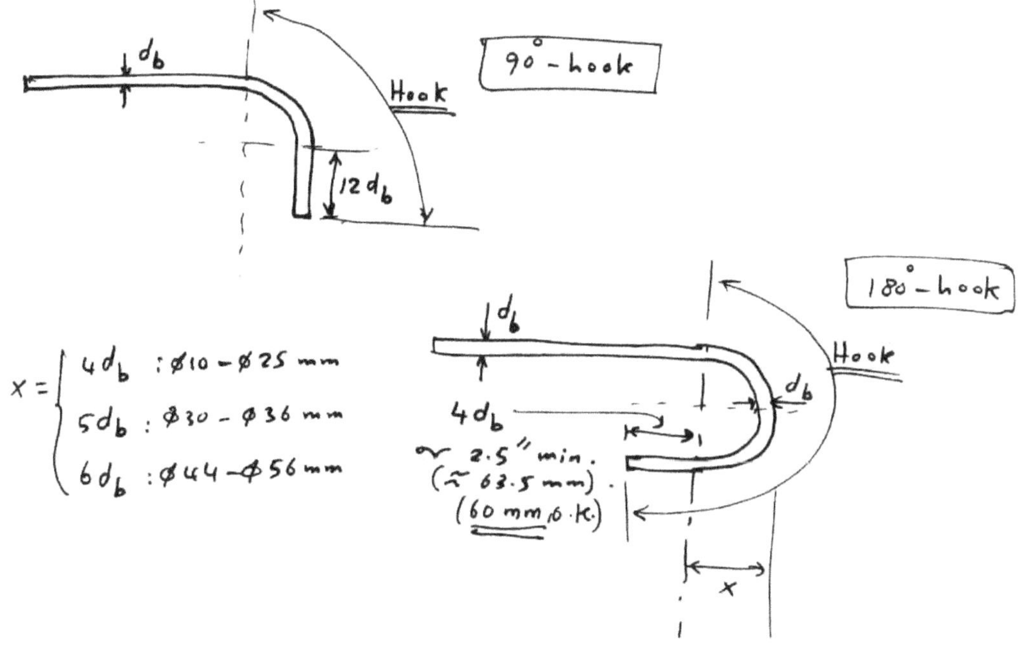

* Bars are typically <u>cold bent</u> to form a hook. However, if cracking occurs during bending, bars may be heated. Maximum temperature not to exceed 1500°F (816°C).

* Tests show that the capacity of a hook increases as the lateral confinement of the concrete within the radius of the hook increases.

* Confinement of the concrete may be raised by:
    - increasing the thickness of the side cover C_s over the hook.
    - increasing the thickness of the top or bottom cover
    - adding closely spaced ties throughout the entire development length of the hook.

* Of these improvements, increasing the side cover and adding stirrups are the most beneficial.

---

* The <u>development length of a hook</u>, ℓ_{dh}, is the horizontal distance that extends from the section of maximum steel stress to the outside face of the hook.

* In the figure below, the section of maximum steel stress occurs at the face of the column (the section of maximum moment) where the flexural stress in the steel is equal to or slightly less than f_y.

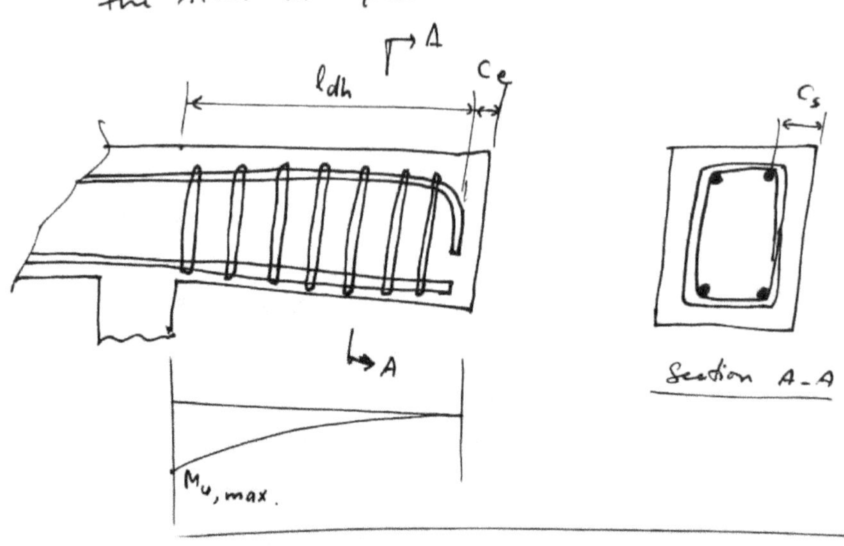

The ACI Design Procedure :

* A bar that terminates in a standard hook is considered safely anchored when the actual horizontal distance between the critical section and the end of the hook equals or exceeds the required development length ℓ_{dh} specified by the ACI Code.

* The required development length of a standard hook is established by multiplying the <u>basic development length</u> ℓ_{hb} by all applicable multipliers listed in Table 6.3 .

The basic development length for a hook represents the anchorage length for a hook:

(1) with less than optimal cover.

(2) with no closely-spaced stirrups to provide confinement

(3) embedded in normal-weight concrete.

* For steel with $F_y = 400\ MPa$, the basic development length is given by ACI Code/Section 12.5.2 :

$$\ell_{hb} = \frac{100\,d_b}{\sqrt{f_c'}} \qquad , \qquad \begin{array}{l} d_b \ \ in \ \ mm \\ f_c' \ \ in \ \ MPa . \end{array}$$

where : d_b : diameter of the hooked bar in mm .

* For effects of thicker cover, lightweight concrete, etc, the required development length of a hook is multiplied by the applicable factors from Table 6.3.

$$\boxed{\ \ell_{dh} = \ell_{hb} * (\text{applicable modifiers from Table 6.3})\ }$$

where $\ell_{dh} \geq 8d_b$ or $150\ mm$, whichever is larger.

Table 6.3 : Modification Factors for l_{hb} :
(Adapted from ACI Code/Section 12.5)

Condition	Modifier
F_y differs from 400 MPa	$\dfrac{f_y}{400}$
Side cover (perpendicular to plane of hook) for $\phi36$ mm bars and smaller is not less than 60 mm. In case of $90°$–hooks, 50 mm or more of end cover must be applied (C_e).	0.7
For $\phi36$ mm and smaller bars, hook enclosed within ties or closed stirrups spaced not more than $3d_b$ of the hooked bar	0.8
Lightweight Concrete	1.3
Epoxy Coating	1.2
$As_{(supplied)} > As_{(required)}$	$\dfrac{As_{(required)}}{As_{(supplied)}}$

<u>Note</u>: Although the ACI Code requires that the development length of top bars be increased by a factor of 1.3 to account for the reduction in bond due to bleeding, the 1.3 factor is <u>not</u> applied to top bars with standard hooks.

* If a bar that terminates at a discontinuous end requires a hook for anchorage, and if the side cover or the top or bottom cover is less than 60 mm, <u>ties</u> must be provided ; they should be at a spacing not to exceed $3d_b$ (d_b is the diameter of the hooked bar) along the entire development length l_{dh}.

* For this situation the 0.8 modifier in Table 6.3 should be omitted.

* Stirrups are not required at the discontinuous ends of the slabs because of the lateral confinement provided by the slab on either side of the hook.

Example 7 :

In the figure shown the top steel in the beam has been designed for a flexural stress f_y at the face of the column, the section of maximum negative moment. If the $\phi 16$ mm bars terminate with a standard $90°$-hook, is sufficient anchorage length available in the column to develop the tensile strength of the steel fully? The side cover on the top bars exceeds 60 mm and the end cover over the vertical leg of the hook equals 60 mm. Take $f_c' = 20$ MPa and $f_y = 400$ MPa.

Solution :

$$l_{hb} = \frac{100 d_b}{\sqrt{f_c'}}$$

$$= \frac{100 (16)}{\sqrt{20}}$$

$$= 358 \text{ mm}.$$

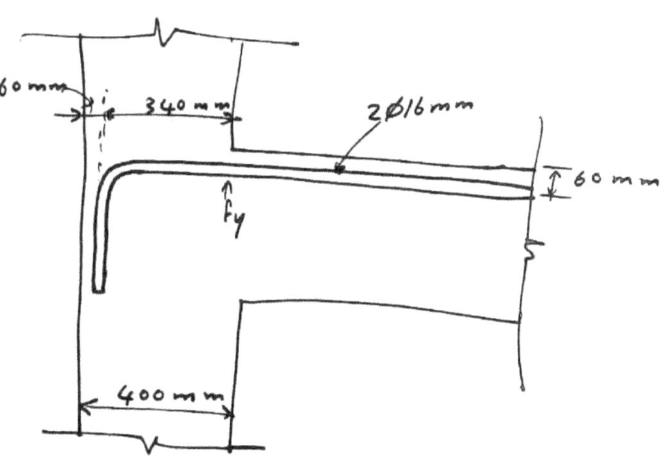

Since the side cover = 60 mm and the end cover of 60 mm exceeds 50 mm, the basic development length may be reduced by 0.7 (see Table 6.3).

Required Development Length $l_{dh} = 358 \times 0.7 = 251$ mm.

Available Development Length $= 340$ mm $> l_{dh} = 251$ mm.

$\Rightarrow$ Anchorage is satisfactory.

Example 8:

Determine the largest-diameter top bar that can be anchored with a $90°$-hook in the 400-mm column shown in the figure of Example 7 if the steel is stressed to f_y at the inside face of the column. 60 mm of side, top, and end cover are provided, and stirrups (not shown in the figure) at a spacing of $3d_b$ will be placed throughout the development length of the hook. Use lightweight concrete with $f_c' = 20$ MPa and $f_y = 345$ MPa.

Solution:

Available hook development length = 340 mm.

$$\ell_{dh} = \frac{100 d_b}{\sqrt{f_c'}} * \text{modifiers} \quad (\text{from Table 6.3}).$$

$$340 = \frac{100 d_b}{\sqrt{20}} \left(\frac{345}{400}\right) (1.3)(0.7)(0.8)$$

where $\frac{345}{400}$ is $\frac{f_y}{400}$.

$$\Rightarrow \quad d_b = 24.2 \text{ mm.}$$

Use a maximum bar size of $\phi 24$ mm.

Example 9:

If $\phi 45$ mm bars in a cantilever beam terminate with a standard $90°$-hook (as shown in the figure), are ACI Code requirements for anchorage satisfactory? $\phi 16$ mm stirrups are spaced 200 mm on center along the length of the hook. The maximum stress in the bar is assumed to develop at the exterior face of the column. $A_{s(required)} = 0.95 A_{s(supplied)}$. Take $f_c' = 36$ MPa and $f_y = 500$ MPa.

Solution :

Basic Development Length :

$$\ell_{hb} = \frac{100 d_b}{\sqrt{F_c'}}$$

$$= \frac{100(45)}{\sqrt{36}}$$

$$= 750 \text{ mm}.$$

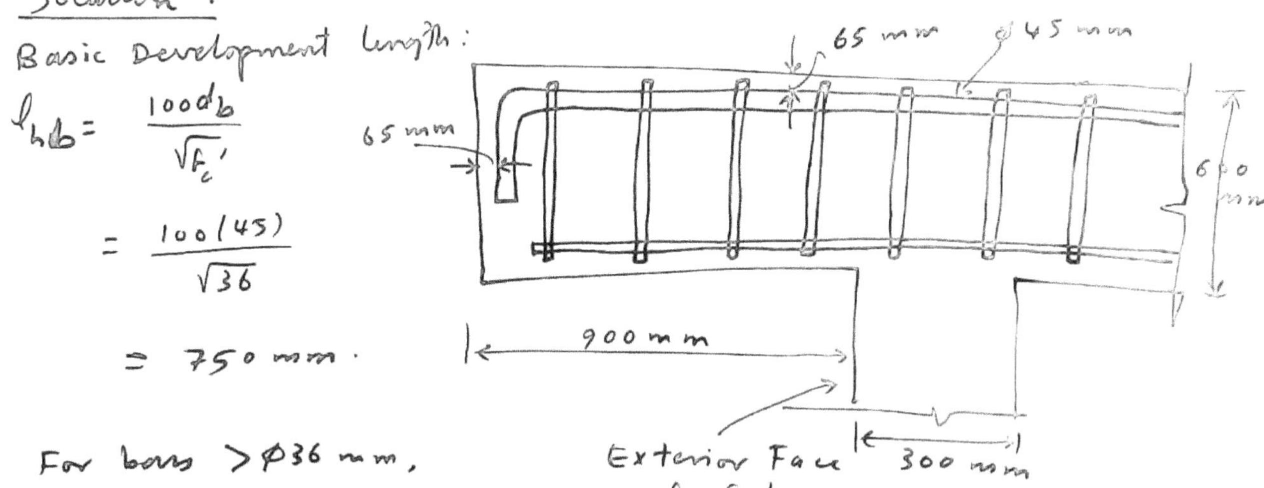

For bars $> \phi 36$ mm, only the reduction for excess steel is allowed.

In addition, we must also increase ℓ_{hb} to account for a yield point in excess of 400 MPa

(see **Table 6.3**)

Required Development Length $\ell_{dh} = \ell_{hb} \left(\frac{As (required)}{As (supplied)} \right) \left(\frac{f_y}{400} \right)$

$$\Rightarrow \ell_{dh} = 750 (0.95) \left(\frac{500}{400} \right) = 891 \text{ mm}.$$

Available development length $= 900 - 65 = 835$ mm.

Since $\ell_{dh} = 891$ mm > 835 mm,

the bar does **not** satisfy ACI Code requirements for anchorage.

6.9 Anchorage of Steel at Simple Supports and Points of Inflection:

* Complete collapse of a beam will occur suddenly if the ends of the positive steel extending into a simple support or into a point of inflection are not properly anchored and slip out.

* Although the moment is zero at these points and the stress in the steel is low, the bond stresses, a function of the rate at which the moment is changing rather than a function of the absolute value of moment, may be large.

* The rate at which the moment changes is related to the shear, which is maximum at a simple support and often high at a point of inflection.

* To prevent a local failure, ACI Code / Section 12.11.3 requires the diameter of the positive reinforcement to be limited so that:

$$\ell_d \leq \begin{cases} \dfrac{M_n}{V_u} + \ell_a & , \text{ at point of inflection} \\[2ex] \dfrac{1.3\,M_n}{V_n} + \ell_a & , \text{ at simple support} \end{cases}$$

where:

V_u : shear at simple support or point of inflection.

M_n : nominal moment strength of section based on area of positive reinforcement extending into support or into point of inflection.

ℓ_a : embedment length beyond centerline of support or beyond point of inflection,

$$\ell_a \leq 12\,d_b \text{ or } d \text{ at point of inflection.}$$

* If a hook is used, it must be converted to an equivalent length of straight bar with the same anchorage capacity.

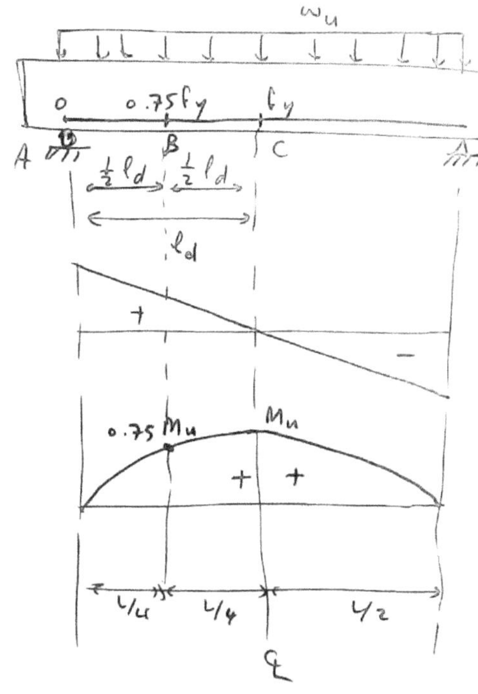

$$M = \frac{w_u L}{2} x - w_u \frac{x^2}{2}$$

$$M = \frac{w_u x}{2}(L - x)$$

S.F.D.

at $x = \frac{L}{2}$,

$$M = \frac{w_u L^2}{8} = M_u .$$

at $x = \frac{L}{4}$,

B.M.D.

$$M = \frac{w_u}{2} \cdot \frac{L}{4} \cdot \frac{3}{4} L$$

$$= \frac{3 w_u L^2}{32}$$

$$= \frac{3}{4}\left(\frac{w_u L^2}{8}\right)$$

$$= \frac{3}{4} M_u .$$

$$= 0.75 M_u$$

In the above example, although the length of the reinforcing bar between the support where the stress is zero and midspan, where the stress equals fy , is equal to the development length, **the bar is <u>not</u> properly anchored.**

At quarter point B, the force in the bar has reduced to $\frac{3}{4} f_y A_b$, a minimum length of embedment of at least $\frac{3}{4} \ell_d$ is required; however, between point B and the support, only a length of $\frac{1}{2} \ell_d$ is available.

$\Rightarrow$ a local bond failure at point B is possible.

Example 10 :

In the shown figure, determine whether the flexural reinforcement (i.e. 3∅28 mm bars) running into the end support satisfies the anchorage requirements of ACI Code (Section 12.11.3 given by the equations on page 204. Bars extend 100 mm past the centerline of supports . Although stirrups are required, they are not shown in side elevation for clarity. In the computation of development length, assume the cover, bar and stirrup spacing (not shown for clarity) permit the designer to select Equation (6.9a) from Table 6.2. Normal weight concrete, $f'_c = 20$ MPa and $f_y = 415$ MPa.

Solution :

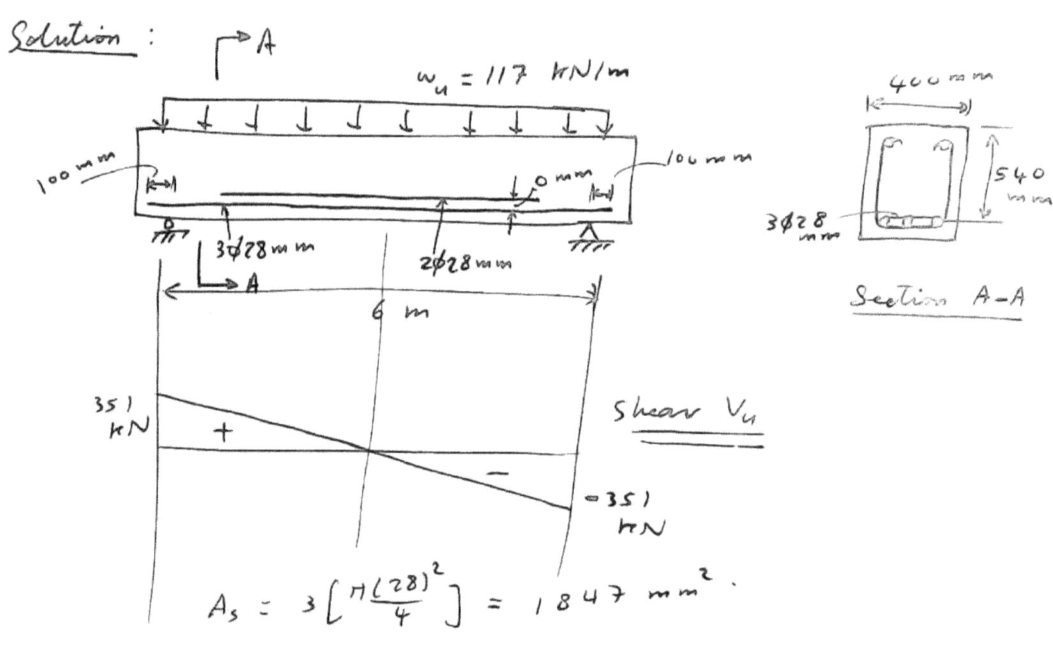

$$A_s = 3\left[\frac{\pi(28)^2}{4}\right] = 1847 \text{ mm}^2.$$

(1) Compute l_d :

$$\frac{l_d}{d_b} = \frac{5 f_y \alpha \beta \lambda}{8\sqrt{f'_c}} = \frac{5(415)(1)(1)(1)}{8\sqrt{20}} = 58.0$$

$$\Rightarrow l_d = 58.0(28) = 1624 \text{ mm}.$$

(2) Verify that $l_d \leq \dfrac{1.3 M_n}{V_u} + l_a$ (for simple supports)

Calculate M_n :

$$T = C$$
$$f_y A_s = 0.85 f'_c a b$$
$$(415)(1847) = 0.85(20) a (400)$$
$$\Rightarrow a = 113 \text{ mm}.$$

$$M_n = T\left(d - \frac{a}{2}\right) = f_y A_s \left(d - \frac{a}{2}\right)$$
$$= (415 \times 1000)(1847)\left[\frac{540}{1000} - \frac{113}{2(1000)}\right] \times 10^{-6}$$
$$= 371 \text{ kN} \cdot \text{m}.$$

$l_a = 100 \ mm \quad$ (given).

$$\frac{1.3 \, M_n}{V_u} + l_a = \frac{1.3 \, (371)}{351} \times 1000 + 100$$

$$= 1474 \ mm \ .$$

(3) Since $l_d = 1624 \ mm > 1474 \ mm$,

reinforcement is **not** properly anchored at the supports.

$$\text{check} \quad l_a \leq \begin{cases} 12 d_b = \text{—} \\ d = \text{—} \end{cases}$$

<u>Note</u> :

If the extension of each $\phi 28 \ mm$ bar running into the end supports is increased from $100 \ mm$ to $300 \ mm$, then, $l_a = 300 \ mm$

and $\quad \dfrac{1.3 \, M_n}{V_u} + l_a = 1674 \ mm$.

$\therefore \quad l_d = 1624 \ mm < 1674 \ mm \implies o.k.$

6.10 Cutoff Points :

* Although using a constant area of reinforcement throughout each region of positive or negative moment is certainly safe, it is often desirable, particularly in heavily reinforced members, to <u>terminate</u> a portion of the steel when the moment decreases significantly (in continuous beams).

* Reducing the area of reinforcement in regions where the moment is low has two major advantages

(1) reduces the volume of reinforcement, the expensive component, thus lowering the cost.

(2) the reduction in congestion in heavily reinforced members produced by terminating excessive bars facilitates the flow of concrete into all regions of the forms and permits improved compaction.

* One or more bars may be terminated at the theoretical cutoff point, at which the moment produced by the factored loads reduces to the moment capacity (i.e. strength) of the cross-section reinforced with the continuing bars.

* Normally the designer terminates 50% to 60% of the steel area sized at the point of maximum moment.

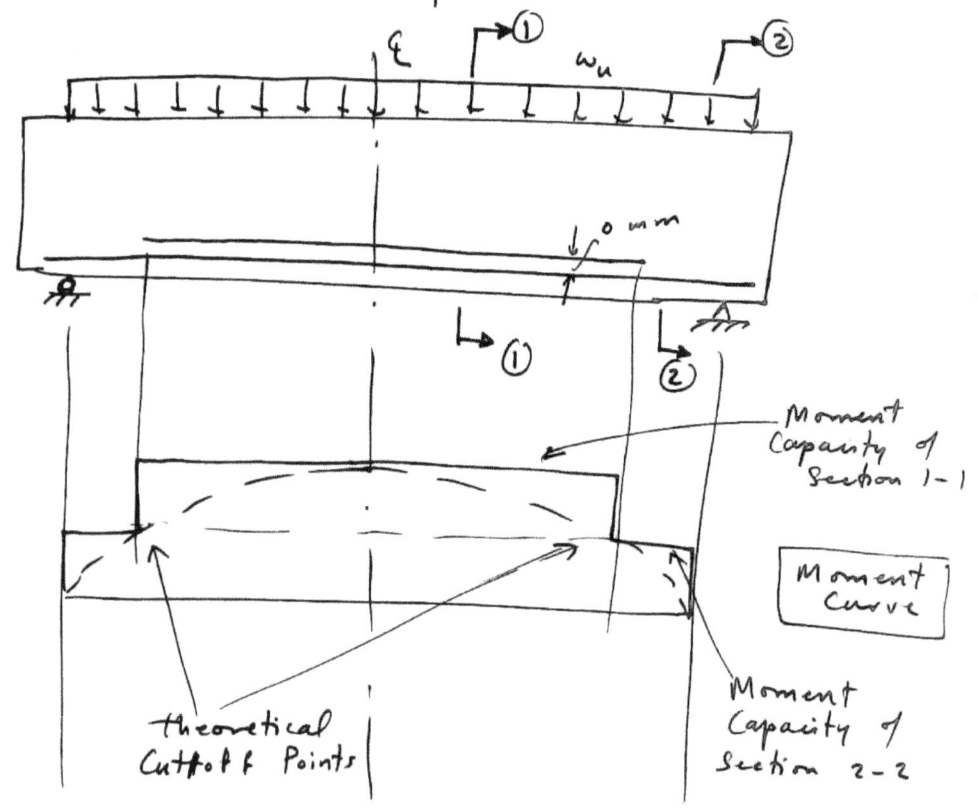

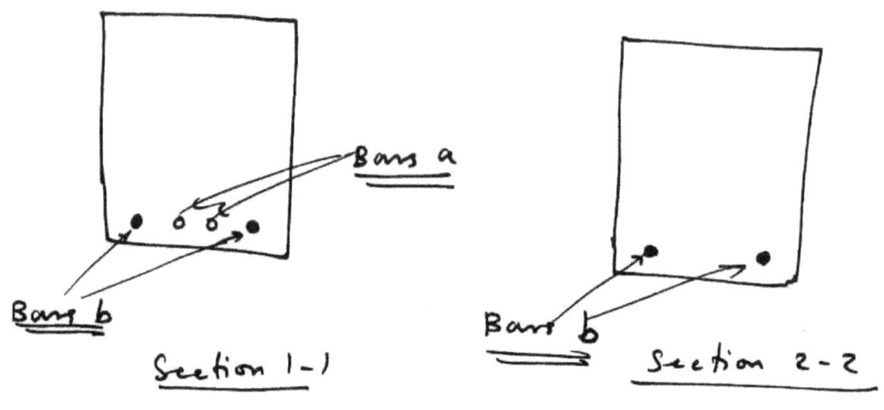

* To account for the possibility of higher-than-anticipated moment at the cutoff point due to possible variations in the position of the live load, settlements of support, or other causes, ACI Code/Section 12.10.3 requires that reinforcement be extended beyond the theoretical cutoff point a distance equal to $12d_b$ or the effective depth d, whichever is larger.

Example 11 :

To carry the midspan moment of $M_u = 317.8$ kN·m, an area of steel of 1465 mm^2 is required in the simply-supported beam shown in the figure. $2\phi22$ mm and $2\phi25$ mm bars, which provide an $A_s = 1742$ mm^2, are selected. Normal weight concrete, $f'_c = 20$ MPa and $f_y = 415$ MPa.

(a) Establish the theoretical cutoff point at which the $\phi22$ mm bars can be terminated.

(b) Considering the extension beyond the cutoff points required by ACI Code/Section 12.10.3, determine the minimum distance the $\phi22$ mm bars should extend on each side of the centerline.

(c) Verify that the $\phi22$ mm bars are properly anchored.

Solution :

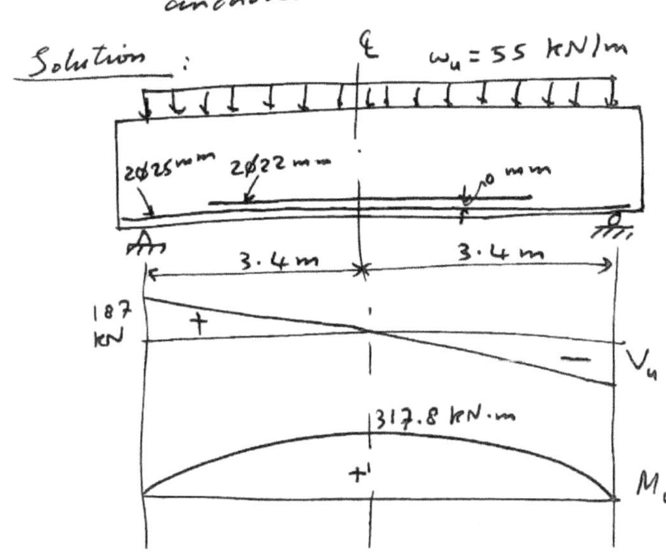

$$M_{u_{max}} = \frac{w_u L^2}{8}$$

$$= 55\,\frac{(6.8)^2}{8}$$

$$= 317.9 \text{ kN·m}$$

(a) To establish the theoretical cutoff point of the $\phi 22$ mm bars, establish the moment capacity of the cross-section reinforced with the two continuing $\phi 25$ mm bars.

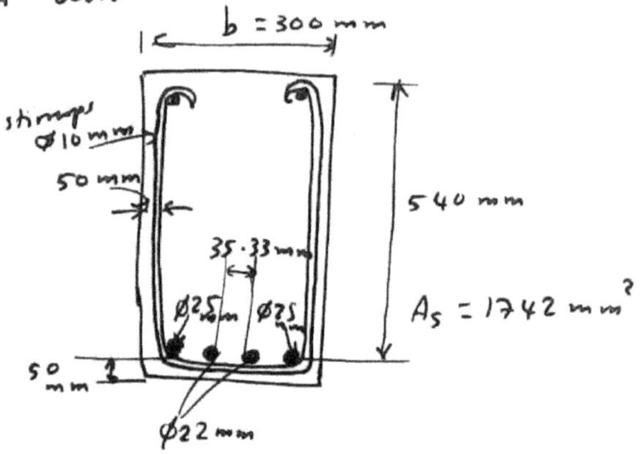

For the $2\phi 25$ mm,

$$A_s = 2\left[\frac{\pi(25)^2}{4}\right] = 982 \text{ mm}^2$$

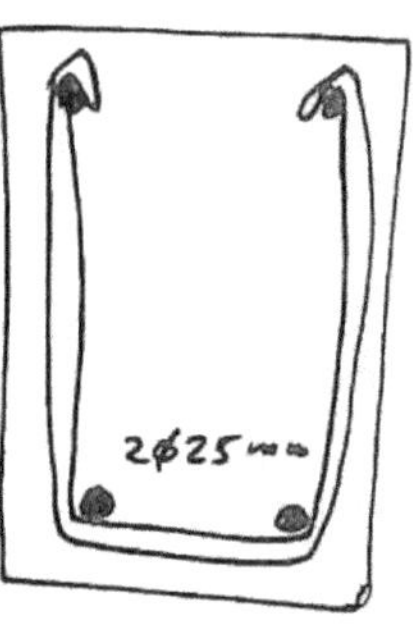

$$C = T$$

$$0.85 f_c' \, ab = f_y A_s$$

$$0.85(20)\,a\,(300) = 415(982)$$

$$\implies a = 79.9 \text{ mm}$$

$$M_n = T\left(d-\frac{a}{2}\right) = f_y A_s \left(d-\frac{a}{2}\right)$$

$$= (415\times10^3)(982\times10^{-6})\left(\frac{540}{1000} - \frac{1}{2}\frac{79.9}{1000}\right)$$

$$= 204 \text{ kN·m}$$

let $M_u = \phi M_n = 0.9(204) = 183.6$ kN·m.

Locate the point on the moment curve at which the moment equals 183.6 kN·m.

$$\frac{x}{3.4} = \frac{y}{187}$$

$$\boxed{y = 55x} \quad \text{or} \quad y = wx$$

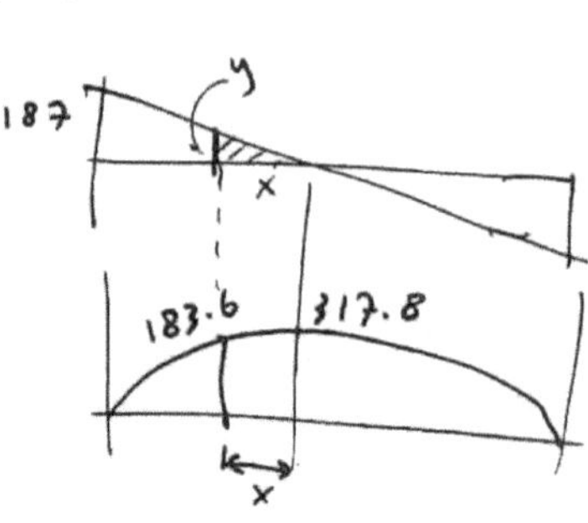

Shaded area under $V = \Delta M$

$$\tfrac{1}{2} x \cdot (55x) = 317.8 - 183.6$$

$$\Rightarrow x = 2.21 \text{ m}. \quad (\text{theoretical cutoff point}).$$

(b) extension beyond theoretical cutoff point:

$$12 d_b = 12 \,(22) = 264 \text{ mm}$$
$$\text{or} \quad d = 540 \text{ mm}$$
$$\Rightarrow \text{take } \underline{540 \text{ mm}}.$$

Length of $\phi 22$ mm bars $= 2.21 + \dfrac{540}{1000} = \underline{2.75 \text{ m}}.$
(from centerline)

$$\text{Total length of } \phi 22 \text{ bars} = 2(2.75) = 5.5 \text{ m}.$$

(c) Check the anchorage of the $\phi 22$ mm bars.

Assume the clear spacing between bars, clear cover, and stirrups (not shown) permit the use of equation (6.99) in Table 6.2.

$$\frac{l_d}{d_b} = \frac{5 f_y \alpha \beta \lambda}{8 \sqrt{f_c'}} = \frac{5\,(415)\,(1)(1)(1)}{8 \sqrt{20}} = 58.0$$

$$\Rightarrow l_d = 58.0\,(22) = 1276 \text{ mm}.$$

Available Anchorage Length $= 2.75 \text{ m} \doteq 2750 \text{ mm}$
$$> 1276 \text{ mm}.$$

Anchorage is o.k.

<u>Notes</u> :

* To ensure that the bending strength of a beam will not be reduced excessively, in regions of small moment the ACI Code limits the area of reinforcement that can be terminated.

* ACI Code/Section 12.11.1 requires that at least one-third of the reinforcement from a section of maximum positive moment be extended (along the face of the member) into a simple support at least 150 mm.

* If a member is continuous, at least one-quarter of the area of reinforcement from the section of maximum positive moment must extend (along the face of the member) into the support at least 150 mm.

* ACI Code/Section 12.12.3 limits the area of negative steel that can be terminated by specifying that at least one-third of the negative reinforcement from the section of maximum moment be extended beyond the point of inflection a distance not less than

    (1) the effective depth d

    (2) $12d_b$ } whichever is larger.

    (3) $\frac{1}{16} \times$ clear span

support at least 6 in or 150 mm (Fig. 6.29).

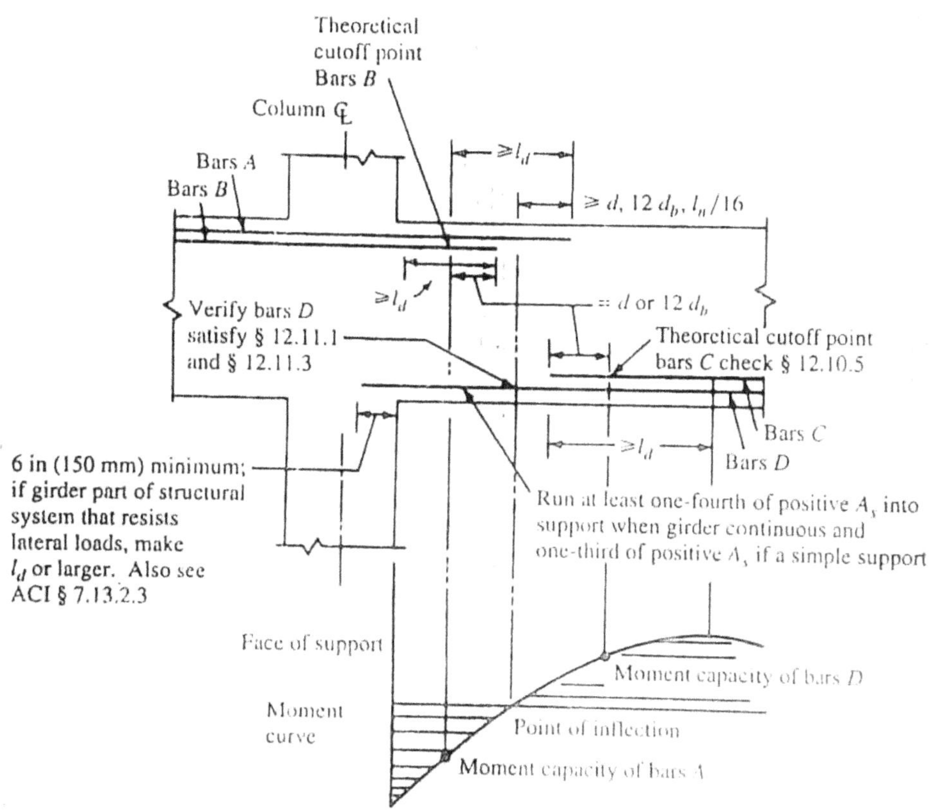

FIGURE 6.29　ACI Code requirements for terminating and anchoring reinforcement.

* At the point in a tension zone where flexural steel is terminated, the stress in the continuing steel increases sharply since a smaller area of steel in the continuing bars must carry the same tension force.

* For the stress in the continuing bars to increase, the beam must undergo a large local increase in strain at the cutoff point. If this increase in strain results in the formation of large tension cracks, the cross-sectional area available to carry shear will be sharply reduced and the possibility of a diagonal-tension failure increased.

* To reduce the likelihood of a diagonal-tension failure developing at a cutoff point, ACI Code/Section 12.10.5 requires that at least one of the following conditions be satisfied:

(1) the shear V_u due to factored loads must not exceed two-thirds of the available shear strength of the cross-section:

$$V_u \leq \frac{2}{3}\phi\left(V_c + V_s\right).$$

(2) For $\phi 36\,mm$ and smaller bars, the area of steel supplied by the continuing bars must be twice that required for moment at the cutoff point, and the shear due to factored loads must not exceed $3/4$ of the shear capacity of the cross-section.

$$A_{s(supplied)} \geq 2\,A_{s(required)}$$

$$V_u \leq \frac{3}{4}\phi\left(V_c + V_s\right)$$

(3) Extra stirrups, in addition to those required for shear and torsion, are supplied. They are to be positioned along each terminated bar for a distance of $\frac{3}{4}d$ from the cutoff point.
the area of the additional stirrups $A_v \geq \dfrac{60 b_w s}{f_y\left(0.4\, b_w s / f_y\right)}$.

In addition, the spacing s of the additional stirrups is not to exceed $\dfrac{d}{8\beta_b}$ where β_b is defined as the ratio of the area of steel terminated to the total area of steel at the cutoff point.

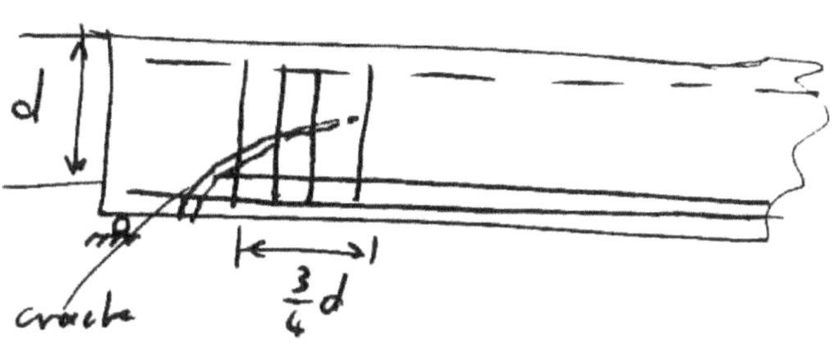

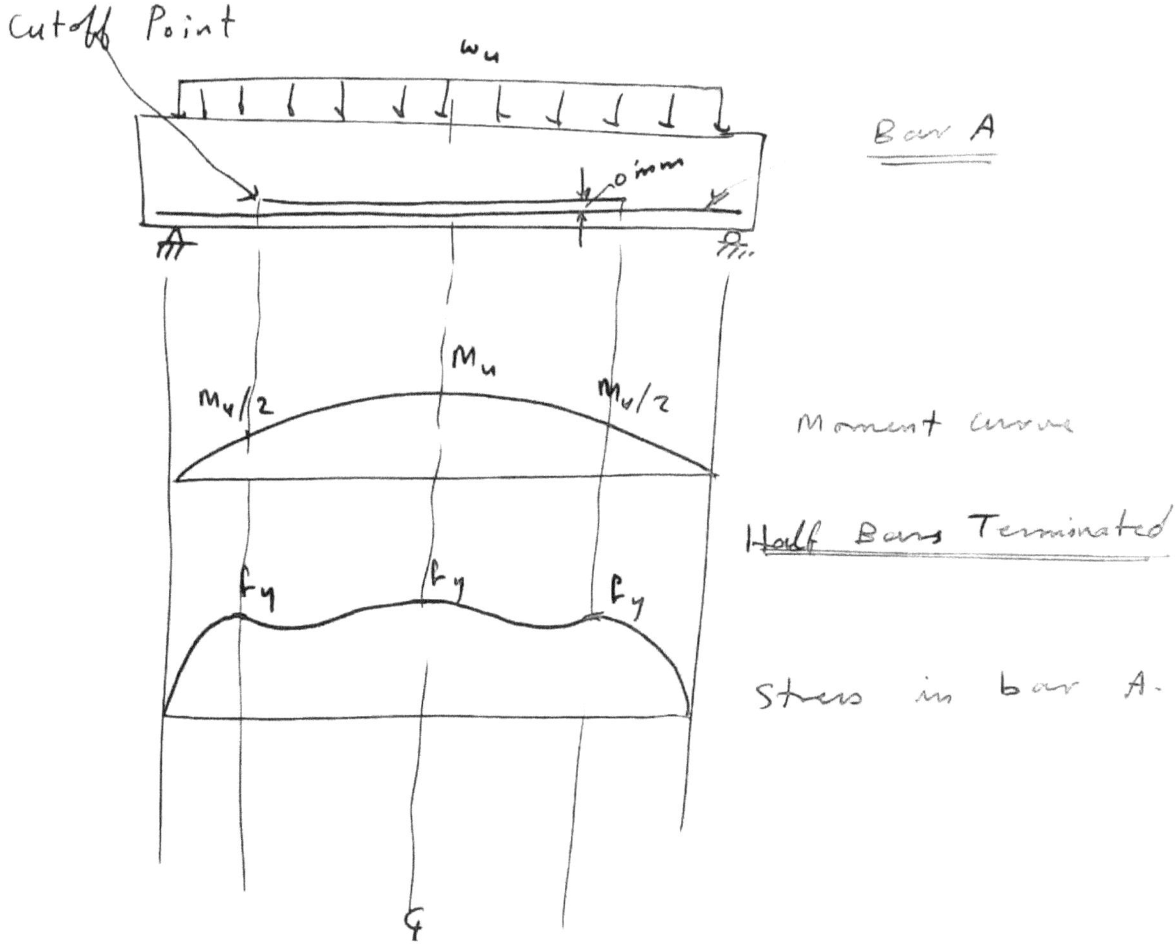

6.11 Splicing Reinforcement:

* Reinforcing bars can be spliced, i.e. made continuous, by welding, by using mechanical connectors, or by lap splices.

* A _lap splice_ is formed by extending bars past each other far enough to permit the force in one bar to be transferred by bond stress through the concrete and into the second bar.

* Although bars joined by a lap splice are usually wired together with their sides in direct contact, forces can also be transferred effectively between bars whose transverse spacing does not exceed $\frac{1}{5} l_s$ or 150 mm, whichever is smaller.

    (l_s : lap length).

* If the reinforcement is heavily stressed at a splice point (the area of reinforcement is less than twice that required to carry the design force), ACI Code (Section 12.15.3 requires the strength of splices made by butt welding or by mechanical connectors to exceed by at least 25% the strength of the reinforcing bar stressed to its yield point.

* Overdesigning the splice ensures that a ductile failure will occur in the unspliced region of the bar by yielding of the steel rather than by rupture of the splice.

* The current edition of the ACI Code permits lap splices of only ϕ 36 mm and smaller bars.

* It is considered good practice to locate splices at sections where stresses are low and to stagger the location of lap splices for individual bars.

* If a splice is defective due to poor workmanship, the strength and ductility of a member will be reduced.

Tension Splices :

* The minimum required lap length of a splice is expressed in terms of development length.

* However, <u>no</u> reduction in development length is allowed when $A_{s(supplied)} > A_{s(required)}$.

* Currently, the ACI Code defines two classes of splices, class A and class B, with minimum lap lengths of l_d and $1.3 l_d$, respectively, but not less than 300 mm.

* The longest lap length is required when the reinforcement is highly stressed and all bars are spliced at the same location.

* ACI Code/Section 12.15.2 requires that lap splices of deformed bars and deformed wires, stressed in tension, be made with class B splices.

* However, a class A splice can be used when the following two conditions are satisfied:

(1) The area of reinforcement supplied is at least twice that required by analysis over the entire length of the splice.

(2) One-half or less of the total reinforcement is spliced within the required lap length.

* When bundled bars are spliced, the lap lengths must be increased to account for the reduced surface of contact between the bars and concrete.

* 20% increase in lap length for a three-bar bundle.
* 33% increase in lap length for a four-bar bundle.
* No overlap of individual bar splices is permitted for bundled bars.

Example 12 : Design of a Tension Splice.

To facilitate construction of a retaining wall, the vertical wall steel shown in the figure is to be spliced to dowels extending from the foundation. If the flexural steel is fully stressed to its yield point at the bottom of the wall, what splice length, ℓ_s is required? Normal weight concrete, $f_c' = 28$ MPa, and $f_y = 415$ MPa.

Solution :

(Note: Horizontal steel is not shown in the figure).

Since 100% of all bars are to be spliced in a region where the steel is fully stressed ($\frac{A_{s(sup.)}}{A_{s(req.)}} = 1$),

a class B splice is required.

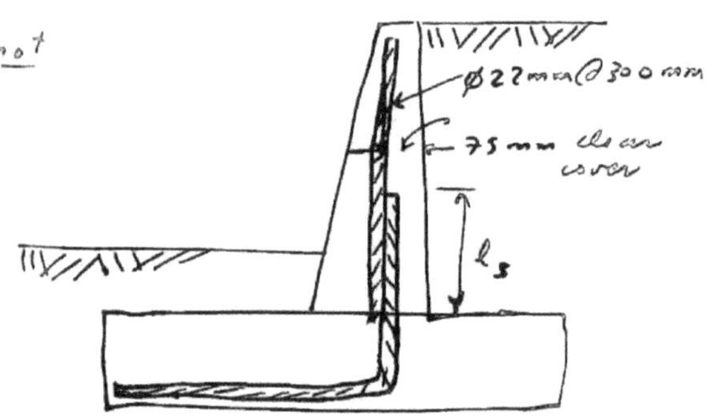

$$\frac{\ell_d}{d_b} = \frac{15 f_y \, \alpha \beta \gamma \lambda}{16\sqrt{f_c'} \left[\frac{c + K_{tr}}{d_b} \right]} \times (0.8)$$

reduction / item (f) in Table 6.1 due to large cover & wide bar spacing

$$c = \begin{cases} \frac{300}{2} = 150 \text{ mm} \\ 75 + \frac{22}{2} = 86 \text{ mm} \end{cases} \Rightarrow \text{Take } c = 86 \text{ mm.}$$

Since no stirrups are used, $A_{tr} = 0 \Rightarrow K_{tr} = 0$

$$\Rightarrow \quad \frac{c + K_{tr}}{d_b} = \frac{86 + 0}{22} = 3.91 \Rightarrow \text{take it to be } \underline{2.5}.$$

From Table 6.1 ,

$\alpha = 1$ (bottom bar)
$\beta = 1$ (uncoated bar)
$\gamma = 1$ (no. 7 & larger).
$\lambda = 1$ (normal weight concrete).

$$\Rightarrow \quad \frac{\ell_d}{d_b} = \frac{15 \, (415) \, (1)(1)(1)(1)(0.8)}{16 \sqrt{28} \, (2.5)} = 23.52$$

$$\ell_d = 23.52 \, d_b = 23.52(22) = 517 \text{ mm}.$$
$$\text{Take } \ell_d = 520 \text{ mm}.$$

$$\text{Splice length (class B)} = 1.3 \, \ell_d = 1.3(520) = 676 \text{ mm}.$$
$$\text{Take } \ell_s = \underline{\underline{680 \text{ mm}}}.$$

Compression Splices :

* For compression lap splices, ACI Code/Section 12.16.1 specifies a minimum lap length equal to :

$$\begin{cases} 0.07 \, f_y \, d_b & \text{for steels with } f_y \leq 400 \text{ MPa} . \\[2mm] (0.13 f_y - 24) \, d_b & \text{for steels with } f_y > 400 \text{ MPa} \end{cases}$$

but not less than 305 mm.

* An additional 33% increase in lap length is required when $f_c' < 20$ MPa.

* When bars of different sizes are lap-spliced in compression, ACI Code/Section 12.16.2 requires that the splice length is to be the <u>larger</u> of :
$$\begin{cases} (1) \text{ the development length of the larger bar.} \\ (2) \text{ the splice length of the small bar.} \end{cases}$$

* Although lap splices are generally prohibited for $d_b > 40$ mm, for compression splices only (i.e. <u>not</u> for tension splices) the Code permits $\phi > 40$ mm bars to be lap-spliced with $\phi 36$ mm and smaller bars.
$$\left[\underline{\text{Note}}: \quad d_b > 40 \text{ mm} \implies \phi 45 \text{ mm} \, \& \, \phi 55 \text{ mm} \right]$$

---

* If the reinforcement in a compression member is confined by <u>ties</u> with an effective area of $\geq 0.0015 \, h_s$ in each direction (where h = dimension of the column perpendicular to the direction of the tie legs, and

s = spacing of the column ties), the length of a [219] lap splice can be reduced by 0.83.

* For reinforcement confined with a _spiral_, the splice length can be reduced by 0.75 because of the increase in strength produced by the confinement of the concrete.

* For both tied and spiral columns, the lap length must be at least 300 mm.

Example 13:

The reinforcement in the column shown in the figure is spliced just above the floor level. If $\phi 22$ mm bars in the upper section of column are spliced to $\phi 25$ mm in the lower section, determine the required length of splice. Take $f_c' = 25$ MPa and $f_y = 415$ MPa.

Solution:

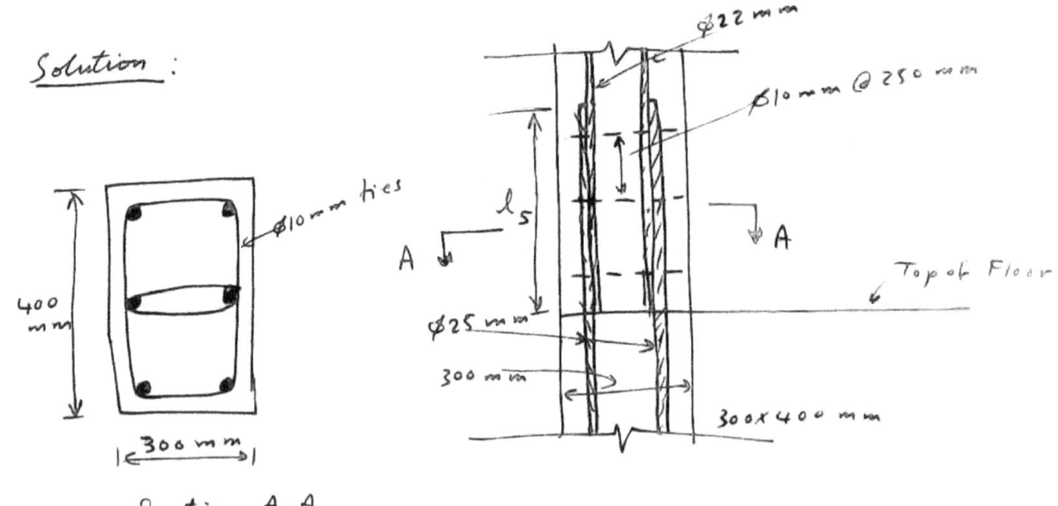

Section A-A

ACI Code / Section 12.16.2:

Minimum splice length is the longer of:
(1) the development length of the longer bar ($\phi 25$ mm).
(2) the splice length of the smaller bar ($\phi 22$ mm).

① For the $\phi 25$ mm bars,

$$l_d = \frac{f_y d_b}{4\sqrt{f_c'}} = \frac{(415)(25)}{4\sqrt{25}} = 519 \text{ mm}.$$

but not less than $0.044\, f_y d_b$

$$= 0.044\,(415)(25) = 457 \text{ mm}.$$

or $200 \text{ mm}.$

② Splice length of the $\phi 22 \text{ mm}$ bars:

$$0.07\, f_y d_b = 0.07\,(415)(22) = 639 \text{ mm}.$$

$$\rightarrow (0.13 f_y - 24)\, d_b = [0.13(415) - 24](22) = 659 \text{ mm}.$$

Use this formula because $f_y > 400 \text{ MPa}.$
but not less than $305 \text{ mm}.$

$\Longrightarrow$ Use 659 mm (controls).

Determine whether the splice can be reduced by 0.83.

A_s for ties in each direction $\geqslant 0.0015\, hs \quad (s = 230 \text{ mm}).$

$\Longrightarrow$ In the long direction, $h = 300 \text{ mm}$ (2 bars),

$$A_s = 2\left[\frac{\pi (10)^2}{4}\right] = 157 \text{ mm}^2.$$

$$A_{s(required)} = 0.0015\, hs = 0.0015\,(300)(250) = 112.5 \text{ mm}^2$$
$$< A_s = 157 \text{ mm}^2.$$

$\Longrightarrow$ In the short direction, $h = 400 \text{ mm}$ (4 bars),

$$A_s = 4\left[\frac{\pi (10)^2}{4}\right] = 314 \text{ mm}^2.$$

$$A_{s(required)} = 0.0015\, hs = 0.0015\,(400)(250) = 150 \text{ mm}^2$$
$$< A_s = 314 \text{ mm}^2.$$

$\Longrightarrow$ Required Splice Length $\ell_s = 0.83\, \ell_d = 0.83\,(659)$
$$= 547 \text{ mm}$$
$$\Longrightarrow \text{Use } \underline{\underline{550 \text{ mm}}}.$$

6.12 Development of Shear Reinforcement:

* For proper function, the shear reinforcement must be carried as close to the compression and tension surfaces of member as cover requirements and proximity of other reinforcement will permit (ACI Code/Section 12.13.1).

* It is especially important to extend the stirrups as close to the compression face as possible because the flexural tension cracks may extend deeply into the compression zone when the nominal strength of the member is approached.

* Hooked stirrups and hooked ties are not permitted for bars larger than $\phi 25\,mm$. (this should rarely apply since stirrup size rarely approaches $\phi 25\,mm$).

* The $180°-$ hook is <u>not</u> used for stirrups and ties.

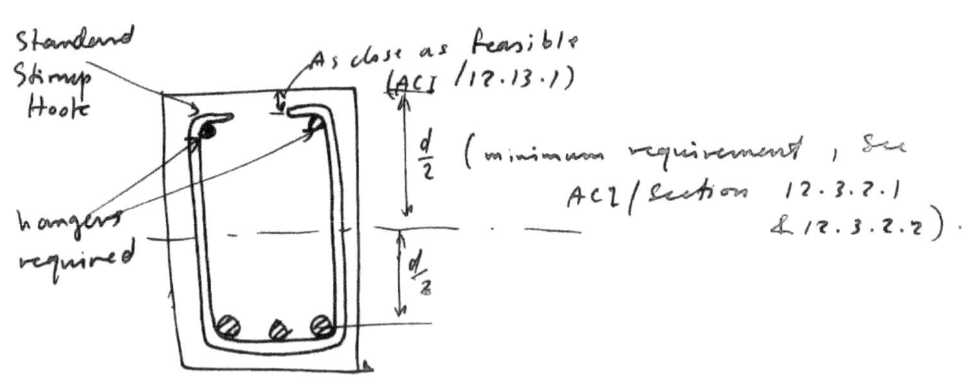

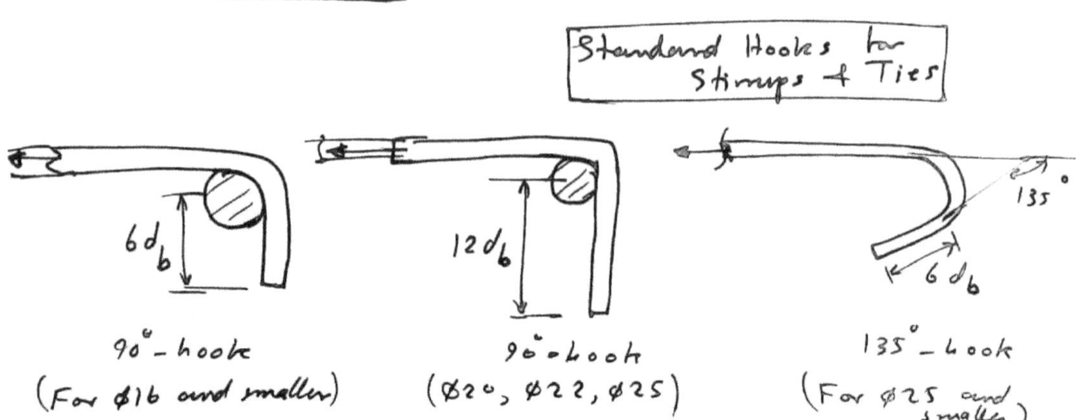

* When <u>closed stirrups</u> are desired, one practical procedure is to use a pair of U-stirrups without hooks placed to form a closed unit. If this is done, ACI/Section 12.13.5 requires laps of $1.3 l_d$ for proper splicing.

* When members are at least 450 mm deep and the tensile capacity $A_b f_y$ of the stirrup does not exceed 40 kN per leg, splices are adequate if the legs extend the full available depth of the member.

Chapter 7

Continuity in Building Frames
of Reinforced Concrete

7.1 Common Building Frames:

* Reinforced concrete building construction commonly has floor slabs, beams, girders, and columns continuously placed to form a monolithic system.

* The complete frame (plan shown in the figure on the next page) should be analyzed as a rigid frame.

* In the analysis for <u>gravity loads</u> on regular building frames, according to ACI / Section 8.9.1 the beams and the adjacent columns may be isolated and treated as a subassembly, with the far ends of the columns assumed as fixed.

* This assumption should not be used for <u>wind loads</u>, however. For wind analysis, except for very tall structures, a simplified approximate method may be used (i.e. the Portal Method, the Cantilever Method). (see Chapter 17 for more details).

* Relative stiffnesses for columns, beams, and girders must first be assumed or established by preliminary design and later reviewed by statically indeterminate structural analysis and design.

7.2 Positions of Live Load For Moment Envelope:

* The live load positions that cause the largest bending moments in slabs, beams, and girders are discussed in this section.

* This is done by constructing qualitative influence lines for the quantities investigated.

* Bending moments to be used in the design of columns are treated separately later.

Section *A–A*

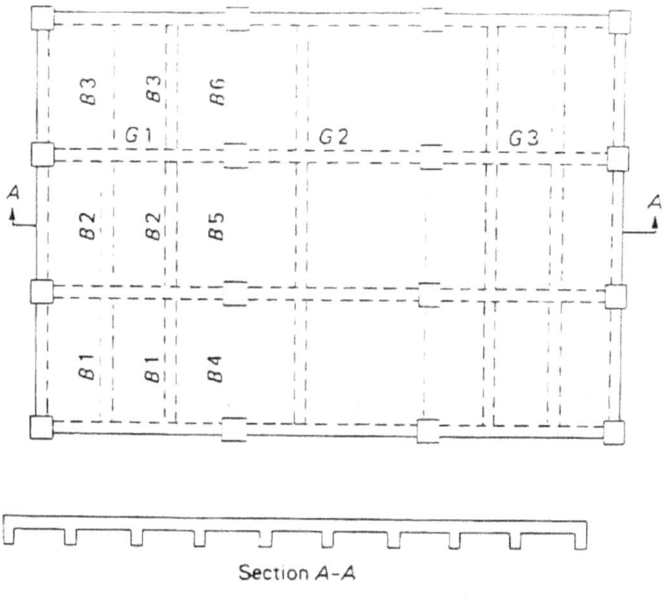

Beams monolithic with girders

Beams monolithic with columns

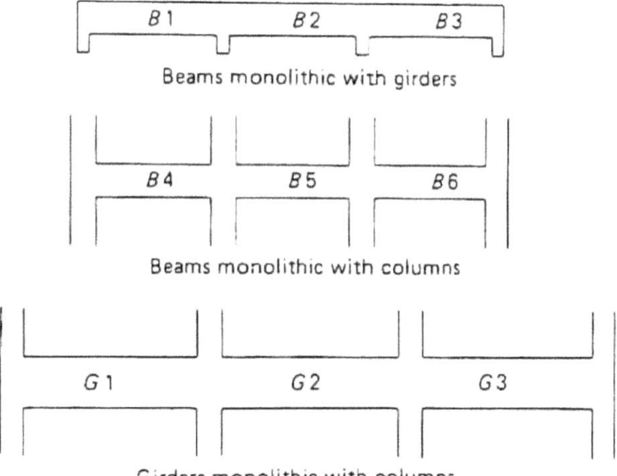

Girders monolithic with columns

Figure 7.1.1 Slab–beam–girder floor

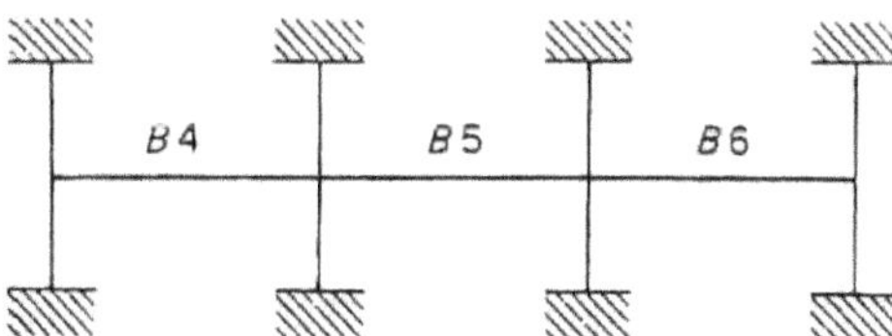

Figure 7.1.2 Beams and adjacent columns

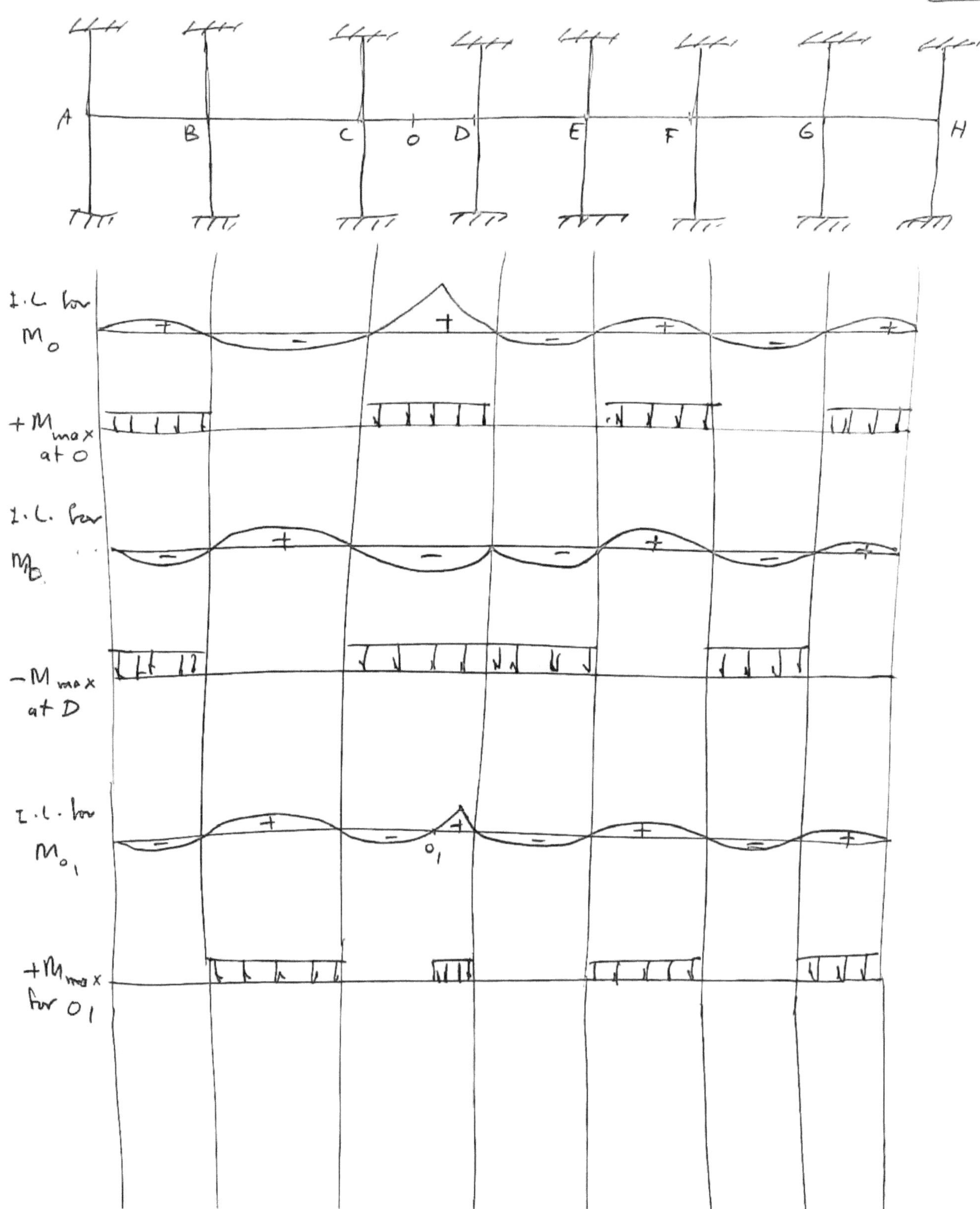
A
B
C
O
D
E
F
G
H
I.L. for M_O
+
−
+
−
+
−
+
$+M_{max}$ at O
I.L. for M_D
−
+
−
−
+
−
$-M_{max}$ at D
I.L. for M_{O_1}
−
+
−
O_1
+
−
+
−
+
$+M_{max}$ for O_1

<u>Rules for Live-Load Placement:</u>

(1) For maximum positive moment within a span, load that span and all other alternate spans.

(2) For maximum negative moment within a span, load the two spans adjacent to that span and all other alternate spans.

(3) For maximum negative moment at a support, load the two spans adjacent to that support and all other alternate spans.

(4) For maximum positive moment at a support, load the two spans beyond each of the two spans adjacent to that support and all other alternate spans.

* From a practical point of view, <u>partial</u> span loading for maximum and minimum bending moments is rarely done in the design of building frames because the effect of doing so on the design is too small to justify the effort.

<u>7.3 Method of Analysis:</u>

* When using the strength method of design, the analysis of the continuous concrete structure is made using <u>factored loads</u>, in accordance with ACI 9.2.

* Use the methods of indeterminate structural analysis.

<u>Example</u>:
Determine the maximum and minimum moments at the middle and ends of each span in a rigid frame with six equal spans as shown in the figure. Assume that the uniform live load w_L is twice the uniform dead load w_D, and that the stiffness factor K_c (representing $\frac{4EI}{L}$) of the column is twice the stiffness factor K_b of the beam span. Express all moments in terms of wL^2 in which $w = w_D + w_L$

and L is the span length.

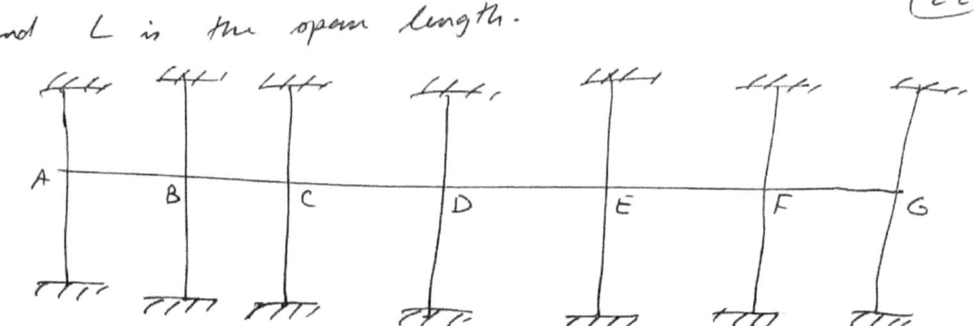

Solution:

* All moments will be expressed numerically in terms of $wL^2 \times 10^{-4}$.

* We have seven loading cases:

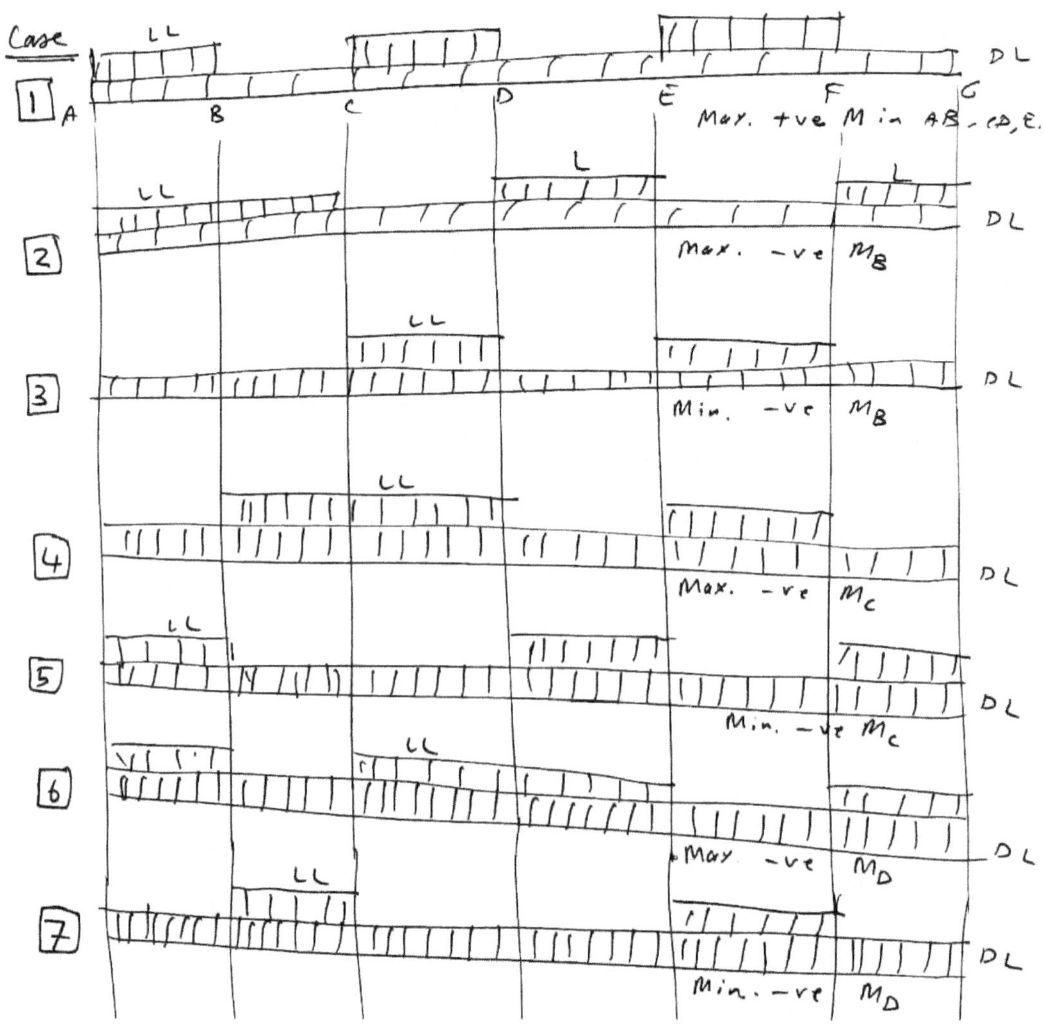

Table 7.3.1   Loading Condition No. 1 (Fig. 7.3.2)

JOINT	A	B		C		D		E		F		G
MEMBER	AB	BA	BC	CB	CD	DC	DE	ED	EF	FE	FG	GF
Distribution factor	$\frac{1}{5}$	$\frac{1}{6}$	$\frac{1}{6}$	$\frac{1}{6}$	$\frac{1}{6}$	$\frac{1}{6}$	$\frac{1}{6}$	$\frac{1}{6}$	$\frac{1}{6}$	$\frac{1}{6}$	$\frac{1}{6}$	$\frac{1}{5}$
FEM	−833	+833	−278	+278	−833	+833	−278	+278	−833	+833	−278	−278
Balance	+167	−92	−92	+92	−92	−92	−92	+92	+92	−92	−92	−56
Carryover	−46	+83	−46	−46	−46	+46	+46	−46	−46	+46	−28	−46
Balance	+9	−22	−22	+15	−15	−15	−15	+15	+15	−3	−3	+9
Carryover	−11	+4	−8	−11	−8	+8	+8	−8	−1	+8	+4	−1
Balance	+2	−2	−2	+3	−3	−3	−3	+1	+1	−2	−2	0
Final M	−712	+804	−340	+331	−777	+777	−334	+332	−772	+790	−399	−184
Change	+121	−29	−62	+53	−56	−56	−56	+54	+61	−43	−121	−94
−½ change	+14	−60	−26	+31	−28	−28	−27	+28	+22	−30	−47	−60
sum	+135	−89	−88	+84	−84	−84	−83	+82	+83	−73	−74	−34
$\theta_{rel} = \text{sum}/K$	+135	−89	−88	+84	−84	−84	−83	+82	+83	−73	−74	−34

Table 7.3.2   Loading Condition No. 2 (Fig. 7.3.2)

MEMBER	AB	BA	BC	CB	CC	DC	DE	ED	EF	FE	FG	GF
FEM	−833	+833	−833	+833	−278	+278	−833	+833	−278	+278	−833	+833
Final M	−668	+911	−888	+729	−325	+337	−778	+777	−331	+340	−804	+712

Table 7.3.3   Loading Condition No. 3 (Fig. 7.3.2)

MEMBER	AB	BA	BC	CB	CD	DC	DE	ED	EF	FE	FG	GF
FEM	−278	+278	−278	+278	−833	+833	−278	+278	−833	+833	−278	+278
Final M	−227	+293	−241	+374	−785	+774	−333	+332	−772	+790	−399	+184

Table 7.3.4 Loading Condition No. 4 (Fig. 7.3.2)

MEMBER	AB	BA	BC	CE	CD	DC	DE	ED	EF	FE	FG	GF
FEM	−278	+278	−833	+833	−833	+833	−278	+278	−833	+833	−278	−278
Final M	−187	+390	−745	+878	−882	+732	−325	−350	−772	−790	−399	−184

Table 7.3.5 Loading Condition No. 5 (Fig. 7.3.2)

MEMBER	AB	BA	BC	CE	CD	DC	DE	ED	EF	FE	FG	GF
FEM	−833	+833	−278	+278	−278	+278	−833	+833	−278	−278	−833	−833
Final M	−709	+813	−385	+225	−228	+379	−786	+774	−330	−340	−804	−712

Table 7.3.6 Loading Condition No. 6 (Fig. 7.3.2)

MEMBER	AB	BA	BC	CE	CD	DC	DE	ED	EF	FE	FG	GF
FEM	−833	+833	−278	+278	−833	+833	−833	+833	−278	+278	−833	+833
Final M	−712	+805	−343	+323	−731	+883	−883	+731	−323	+343	−805	−712

Table 7.3.7 Loading Condition No. 7 (Fig. 7.3.2)

MEMBER	AB	BA	BC	CE	CD	DC	DE	ED	EF	FE	FG	GF
FEM	−278	+278	−833	+833	−278	+278	−278	+278	−833	+833	−278	+278
Final M	−184	+398	−787	+780	−378	+228	−228	+378	−780	+787	−398	+184

7.3.7 and entered in Table 7.3.8. Note that the designer's sign convention for bending moment. in which a positive moment causes compression on the top side of the beam. is used in Table 7.3.8.

The moment at the midspan M_s may be determined by superposition of the effect of end moments with that of the simple beam moment due to transverse loading.

$$M_s = M_0 - \tfrac{1}{2}(M_L + M_R)$$

in which M_0 is the moment at the midspan for a simple beam. When the end moments are not equal, the maximum moment in the span does not occur at midspan. but its value is close to that at midspan.

Table 7.3.8   Summary of Results of Moment Distribution Analysis

LINE NUMBER		SPAN	M_L	M_R	VALUES OF M_L AND M_R FROM TABLE NO	$M_s = M_0 - \frac{1}{2}(M_L + M_R)$
1	For maximum positive moment at midspan	AB	-712	-804	7.3.1	$+492$
2		BC	-790	-772	7.3.1	$+469\ (M_0 = +1250)$
3		CD	-777	-777	7.3.1	$+473$
4	For minimum positive or maximum negative moment at midspan	AB	-184	-399	7.3.1	$+125$
5		BC	-340	-331	7.3.1	$+81\ (M_0 = +417)$
6		CD	-332	-334	7.3.1	$+84$
7	For maximum negative moment at	A of AB	-712	(-804)	7.3.1	
8		B of AB	(-668)	-911	7.3.2	
9		B of BC	-888	(-729)	7.3.2	
10		C of BC	(-745)	-878	7.3.4	
11		C of CD	-882	(-732)	7.3.4	
12		D of CD	(-731)	-883	7.3.6	
13	For minimum negative or maximum positive moment at	A of AB	-184		7.3.1	
14		B of AB		-293	7.3.3	
15		B of BC	-241		7.3.3	
16		C of BC		-225	7.3.5	
17		C of CD	-228		7.3.5	
18		D of CD		-228	7.3.7	

Note: Values of moments in $wL^2(10^{-4})$. Numbers in parenthesis are to be used in shear envelope computations (see Section 7.6).

For the purpose of illustration, the moment diagram for the maximum and minimum positive moments at midspan, using results of the first loading condition in Fig. 7.3.2, is shown in Fig. 7.3.3. First, the simple beam moment diagrams for the total load and for dead load only are drawn to scale in Fig. 7.3.3(a). Next, the end moments for each span are taken from Table 7.3.8 and plotted in Fig. 7.3.3(b). Note that, because of symmetry, only the moment diagrams for the first three spans are shown. The final moment diagrams in Figure 7.3.3(b) are drawn by rotating the base lines for zero end moments in Fig. 7.3.3(a) to those connecting the end moments in Fig. 7.3.3(b).

In this example, the dead load has been applied on all the spans in each of the seven loading conditions of Fig. 7.3.2. This has been done because an important purpose of the example is to justify the use of approximate moment coefficients as discussed in the next section. An alternative approach would be to use eight loading approximately equal spans (the larger of two adjacent spans not exceeding the shorter by more than 20%) with loads uniformly distributed, where the unit live load does not exceed three times the unit dead load. One approximate method would be the method of isolating one floor at a time (with its upper and lower columns) as permitted by ACI-8.9.1 and discussed in Section 7.3. The ACI Code leaves open to the designer the precise definition of the term "approximate methods." The reader is referred to the subassembly model defined in Fig. 7.1.2.

It seems desirable to examine the moment coefficients in ACI-8.3.3. Observation of Table 7.3.8 in the preceding section shows that for a six-span frame in which the ratio of ΣK_{col} to K_{bm} is 4 and the ratio of w_L to w_D is 2, critical values of moments may vary within the following limits:

Exterior span:	
Exterior end	$-0.0184wL^2$ and $-0.0712wL^2$
Midspan	$+0.0125wL^2$ and $+0.0492wL^2$
Interior end	$-0.0293wL^2$ and $-0.0911wL^2$
First interior span:	
Exterior end	$-0.0241wL^2$ and $-0.0888wL^2$
Midspan	$+0.0081wL^2$ and $+0.0469wL^2$
Interior end	$-0.0225wL^2$ and $-0.0878wL^2$
Second interior span:	
Exterior end	$-0.0228wL^2$ and $-0.0882wL^2$
Midspan	$+0.0084wL^2$ and $+0.0473wL^2$
Interior end	$-0.0228wL^2$ and $-0.0883wL^2$

Similar values may be worked out for other values of $(\Sigma K_{col})/K_{bm}$ and of w_L/w_D. It may be observed that the maximum positive moments in the first and second interior spans are about equal, that the maximum positive moment in the exterior span is higher than that in the interior spans, that the maximum negative moment at the interior end of the exterior span has the largest numerical value, and that the maximum negative moments at both ends of all interior spans are about equal. The moment coefficients in ACI-8.3.3 are in agreement with these observations.

As a matter of further justification of the ACI moment coefficients, a comparison of these values with the largest possible theoretical values [7.4] is shown in Table 7.4.1. Certainly the largest possible theoretical values will be for the case of $w_L/w_D = 3$, which is the limit set forth in ACI-8.3.3. In this instance, secondary live load moments with signs opposite to that of dead load occur infrequently; or, if they do occur, their values are small. Thus, as long as the ratio of live load to dead load is well within 3 and span lengths do not differ considerably, there will be no moment reversal so that the ACI moment coefficients are reasonably close and, in general, on the safe side. It may be noted, however, that the ACI moment coefficients are in terms of wL_n^2, in which L_n is the clear span for positive moment and the average of the two adjacent clear spans for negative moment—negative moments being those at the face of supports and not at the centerline of support. On the other hand, the theoretical coeffi-

* The analysis will give the moments at the joints:

M_s : midspan moment

M_0 : midspan moment for a simply-supported beam

$$\left(M_0 = \tfrac{1}{8} w L^2 \right).$$

$$\Longrightarrow \boxed{ M_s = M_0 - \tfrac{1}{2} \left(M_L + M_R \right) }$$

<u>Note</u>: When the end moments are <u>not</u> equal (i.e. $M_L \neq M_R$), the maximum moment in the span does not occur at midspan, but its value is close to that at midspan.

<u>Note</u>: See next pages (Table $\underline{7.3.8}$ for the results).

7.4 ACI Moment Coefficients :

* ACI/section 8.3 specifies that:

(1) the theory of elastic analysis is to be used in analyzing frames or continuous construction,

(2) except for prestressed concrete, approximate methods of frame analysis are permitted for buildings of <u>usual</u> types of construction, spans, and story heights.

(3) except for prestressed concrete, design for the moments and shears as listed in ACI/section 8.3.3 is satisfactory in the case of two or more approximately equal spans (the larger of two adjacent spans not exceeding the shorter by more than 20%) with loads uniformly distributed, where the unit live load does not exceed three times the unit dead load.

Table 7.4.1   Comparison of ACI Moment Coefficients with Theoretical Values [7.4]

LOCATION OF SECTION	ACI	THEORETICAL COEFFICIENTS			
		VALUE	NUMBER OF SPANS	$(\Sigma K_{col})/K_{bm}$	w_L/w_D
Positive moment					
End spans					
If discontinuous end					
is unrestrained	$-\frac{1}{11}$	$+0.094$	3	0	3
If discontinuous end is					
integral with the					
support	$-\frac{1}{14}$	$+0.073$	3	0.5	3
Interior spans	$-\frac{1}{16}$	$+0.063$	4 or more	0.5	3
Negative moment at					
exterior face of first					
interior support					
Two spans	$-\frac{1}{9}$	-0.111	2	0.5	3
More than two spans	$-\frac{1}{10}$	-0.107	4 or more	0.5	3
Negative moment at other					
faces of interior					
supports	$-\frac{1}{11}$	-0.092	4 or more	2	3
Negative moment at face					
of all supports for (a)					
slabs with spans not					
exceeding 10 ft and (b)					
beams and girders					
where $(\Sigma K_{col})/K_{bm}$					
exceeds 8 at each end of					
the span	$-\frac{1}{12}$	-0.083	any number	∞	any ratio
Negative moment at					
interior faces of exterior					
supports for members					
built integrally with					
their supports					
Where the support is					
a spandrel beam or					
girder	$-\frac{1}{24}$	-0.036	4 or more	0.5	3
		-0.050	4 or more	1	3
Where the support is					
a column	$-\frac{1}{16}$	-0.064	4 or more	2	3

cients are in terms of wL^2, in which L is the distance between centerlines of supports. and coefficients for negative moments refer to those at the centerlines of support. Although span lengths between centerlines of supports are always used in elastic analysis, ACI-8.7.3 states that for beams built integrally with supports, moments at faces of supports may be used for design.

Observation on Moment Coefficients of ACI/8.3.3 :

(1) The maximum positive moments in the first and second interior spans are about equal.

(2) The maximum positive moment in the exterior span is higher than that in the interior spans.

(3) The maximum negative moment at the interior end of the exterior span has the largest numerical value.

(4) The maximum negative moments at both ends of all interior spans are about equal.

Note : The ACI moment coefficients are in terms of wL_n^2 where

$$L_n = \begin{cases} \text{clear span for positive moment} \\ \text{average of two adjacent clear spans for negative moment} \end{cases}$$

* Negative moments are given at the face of supports and not at the centerline of support.

* ~~theoretical~~ values given (from the Example) are in terms of wL^2, where L is the distance between centerlines of supports, and coefficients for negative moments refer to those at centerlines of support.

* Although span lengths between centerlines of supports are always used in elastic analysis, ACI/Section 8.7.3 states that for beams built integrally with supports, moments at faces of supports may be used for design.

7.5 ACI Moment Diagrams:

* Two primary sets of shear and moment diagrams are inherently assumed.

(1) loading position that causes maximum positive moment within the span.

Logically, any approximate analysis that provides adequate strength and no excessive deflection at service load should be acceptable. Furlong and Rezende [7.5] have proposed an alternative to the use of ACI-8.3. The alternate is based on limit analysis as discussed in Section 10.12, with requirements that every component be ductile and strong enough, and that no reinforcement yields at service load.

7.5 ACI MOMENT DIAGRAMS

In designing any span in a multispan continuous rigid frame subjected to live load where moment coefficients are used, two primary sets of shear and moment diagrams are inherently being assumed. In the general case, one will result from the loading position that causes maximum positive moment within the span, and the other will result from assuming that the maximum negative moments occur simultaneously at both ends. Actually, the loading position that causes maximum negative moment at

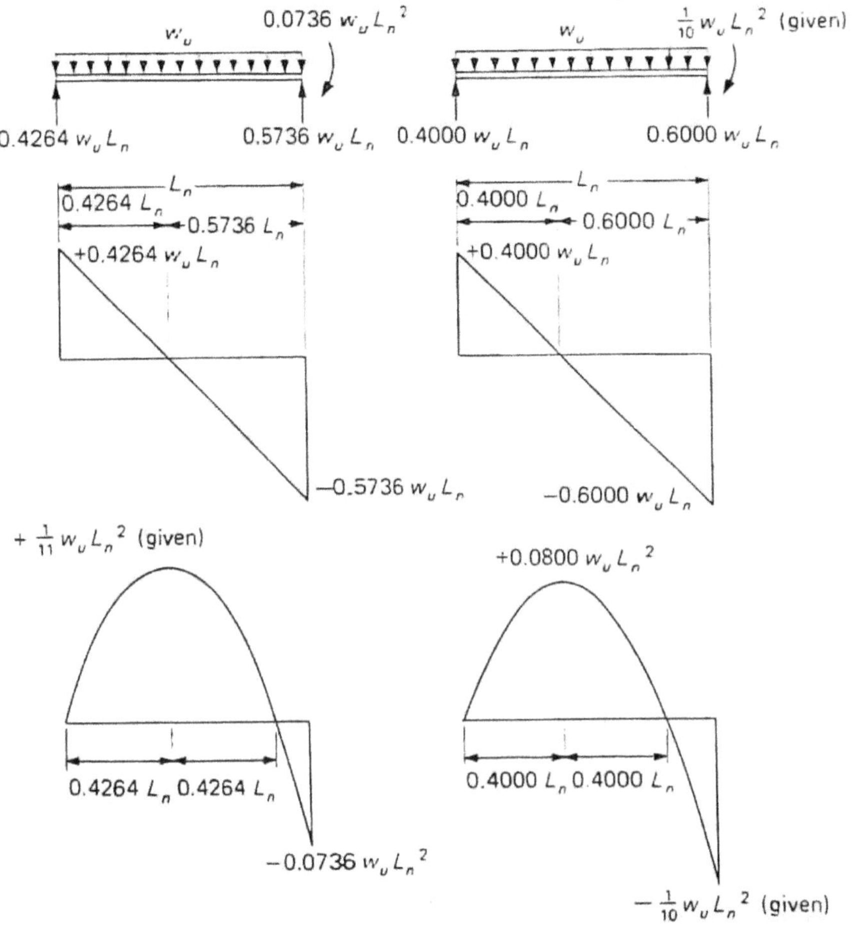

(a) Maximum in the positive zone   (b) Maximum in the negative zone

Figure 7.5.1 Exterior span with discontinuous end unrestrained

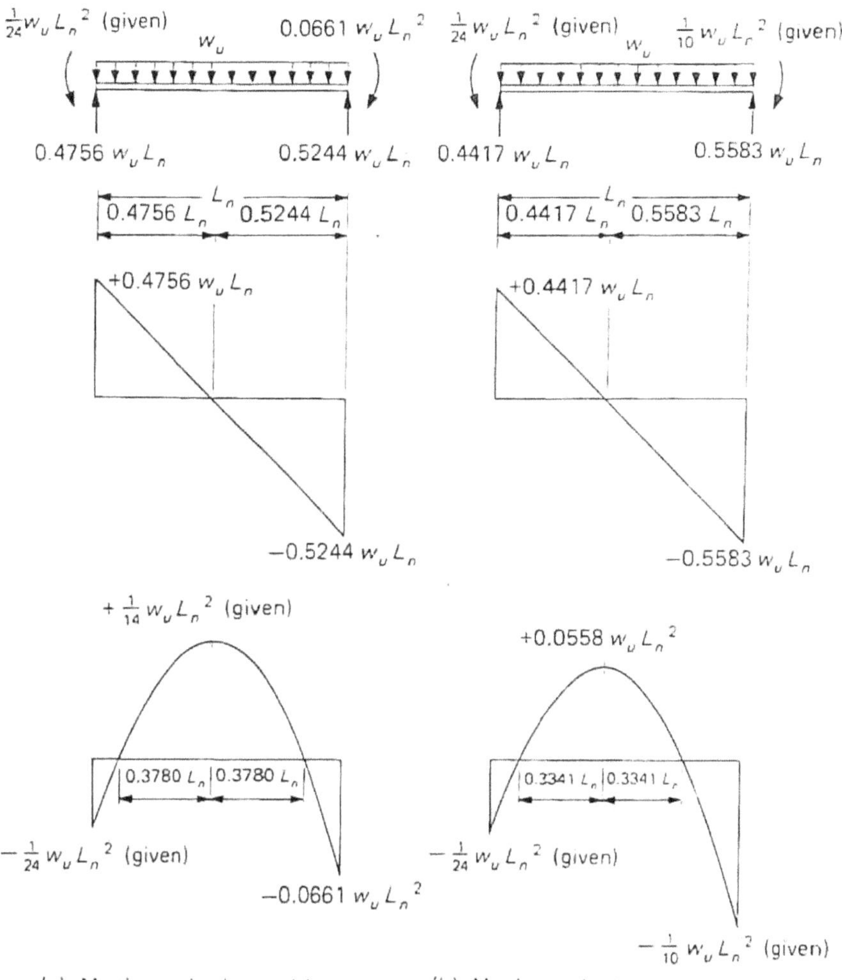

Figure 7.5.2 Exterior span with exterior support built integrally with spandrel beam or girder

one end is different from that which causes maximum negative moment at the other end. However, by assuming that both maximum negative end moments occur simultaneously, a critical curve having greater magnitude than either of the two actual curves is obtained.

The ACI moment coefficients (ACI-8.3.3) shown in Table 7.4.1 are the common values from the two primary conditions as described in the preceding paragraph. No secondary moment coefficients are suggested by the Code, the reason being that as long as the design live to dead load ratio is limited to 3, no moment reversal will occur; that is, there can be only positive moment in the midspan region and only negative moment in the support region.

In Figs. 7.5.1 to 7.5.4, inclusive, are shown the two primary sets of shear and moment diagrams, for the various conditions, to be used in the design of continuous spans in accordance with the ACI moment coefficients. Note that these diagrams are

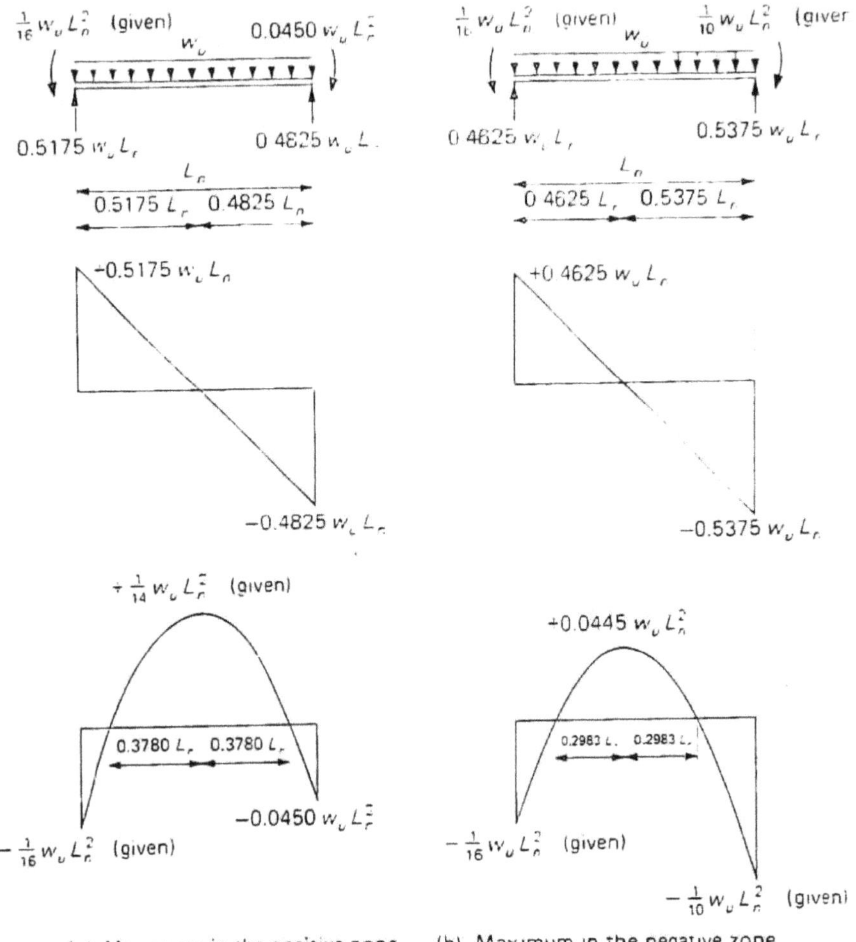

Figure 7.5.3 Exterior span with exterior support built integrally with column.

applicable to the actual *clear* span, which is also used to compute the positive moment. For negative moment, L_n is the average of adjacent clear spans (ACI-8.0).

The reader should utilize the fundamentals of shear and moment diagrams to verify the numerical ordinates on these diagrams. For instance, in case of maximum positive moment in Fig. 7.5.2(a), the distance x from the left support to the point of zero shear may be determined from the relationship that the change of moment between any two sections is equal to the area of the shear diagram between these two sections. Thus

$$\frac{w_u x^2}{2} = \left(\frac{1}{24} + \frac{1}{14}\right) w_n L_n^2$$

from which

$$x = 0.4756 L_n$$

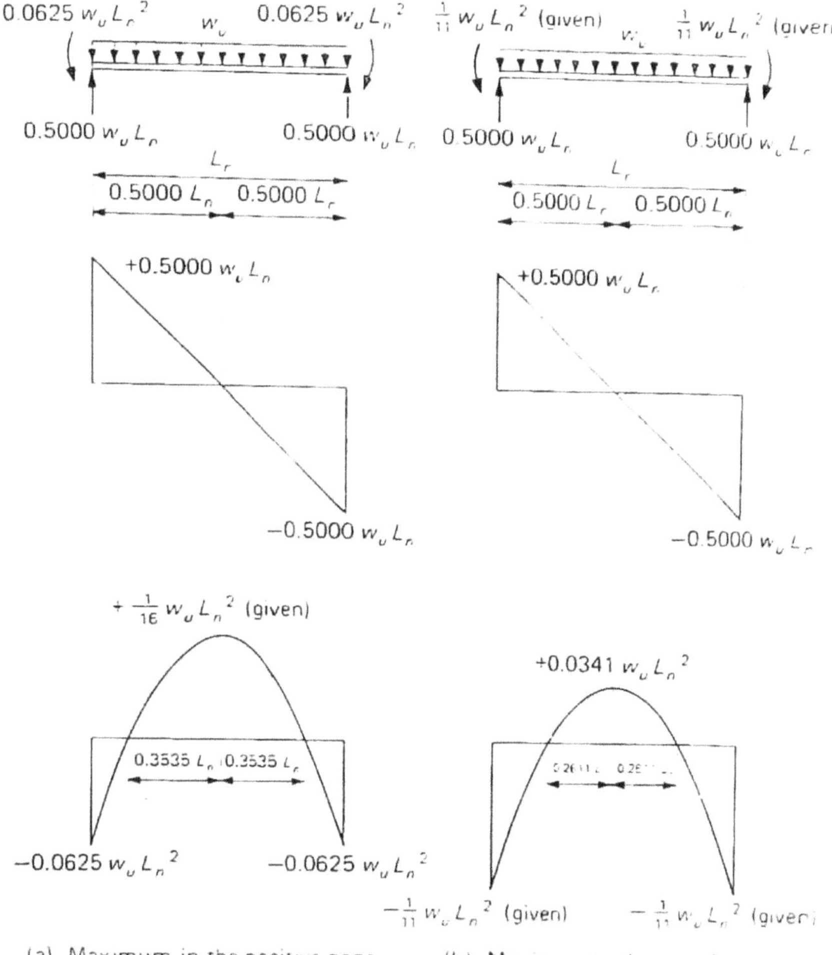

Figure 7.5.4 Interior span

Also, the distance x between the section of maximum positive moment to the point of zero moment is

$$\frac{w_u x^2}{2} = \frac{1}{14} w_n L_n^2$$

from which

$$x = 0.3780 L_n$$

Any time a designer uses moment coefficients for determining the factored moments, as permitted by the approximate method of ACI-8.3.3, the moment diagrams that correspond to such coefficients should be used when establishing bar bend or cutoff locations. The use of a given moment coefficient implies a statically compatible moment diagram.

(2) maximum negative moments occur simultaneously at both ends.

Actually, the loading position that causes maximum negative moment at one end is different from that which causes maximum negative moment at the other end.

However, by assuming that both maximum negative end moments occur simultaneously, a critical curve having greater magnitude than either of the two actual curves is obtained.

Note: As long as the design live to dead load ratio is limited to 3, no moment reversal will occur; that is, there can be only positive moment in the midspan region and only negative moment in the support regions.

---

* the two primary sets of shear and moment diagrams, for the various conditions, to be used in the design of continuous beams, are shown on the next four pages.

Check some values:

① In Figure 7.5.2(a):

To locate point 0 of zero shear:

$$M_0 = M_L + \text{area of shear diagram}$$

$$\frac{1}{14} w_u L_n^2 = -\frac{1}{24} w_u L_n^2 + w_u x \left(\frac{x}{2}\right)$$

$$\frac{1}{2} w_u x^2 = \left(\frac{1}{14} + \frac{1}{24}\right) w_u L_n^2$$

$$\implies x = \sqrt{2\left(\frac{1}{14} + \frac{1}{24}\right)} \cdot L_n$$

$$\implies \boxed{x = 0.4756 \, L_n}$$

(2) In Figure 7.5.2(a),

To locate point P of zero moment:

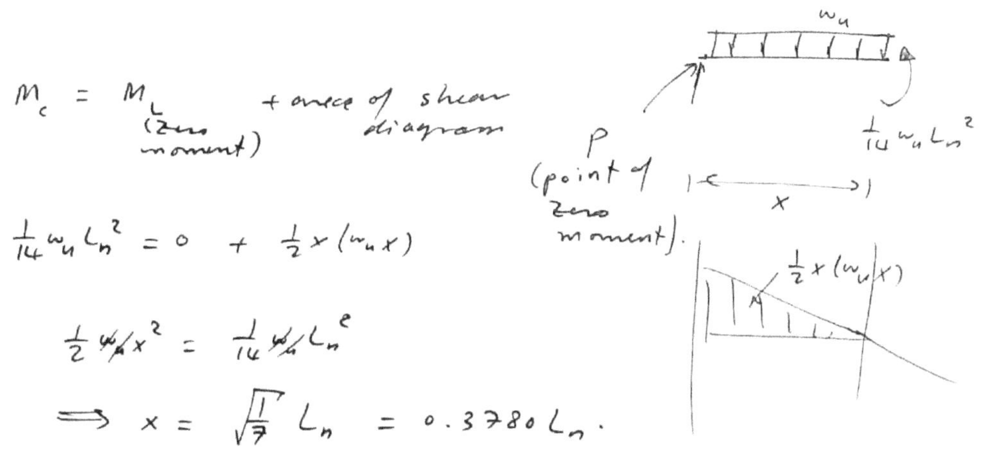

$$M_c = M_{L\,(\text{zero moment})} + \text{area of shear diagram}$$

$$\tfrac{1}{14} w_u L_n^2 = 0 + \tfrac{1}{2} x (w_u x)$$

$$\tfrac{1}{2} w_u x^2 = \tfrac{1}{14} w_u L_n^2$$

$$\Rightarrow x = \sqrt{\tfrac{1}{7}}\, L_n = 0.3780\, L_n.$$

7.6 Shear Envelope for Design:

* The proper position of live load for maximum shear at a section can be determined by examining the influence line for shear at that section along the span.

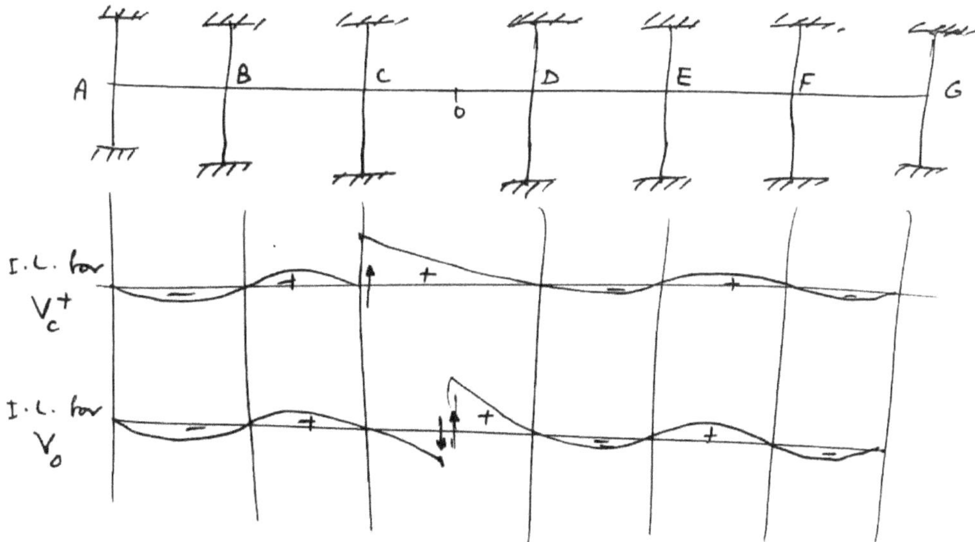

* Partial span loading should be used.

However, when ACI moment coefficients are used, it is generally assumed that the shear diagrams accompanying the critical moment diagram may be used in the design.

23

Chapter 8

Design of One-Way Slabs

8.1 Introduction:

* For a slab, when the long direction is twice as long as the short direction,

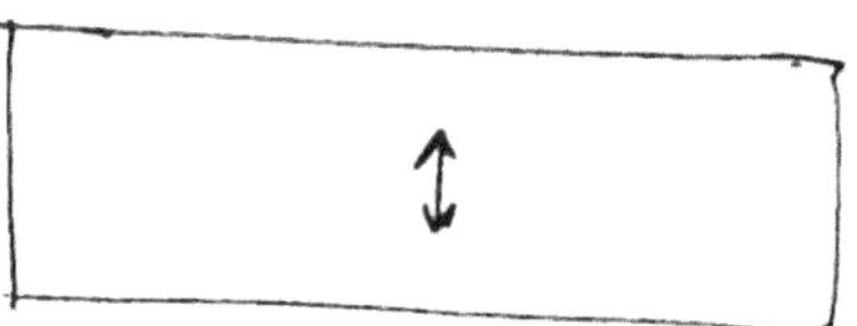

  the dead and live load acting on the slab area may be considered as being entirely supported in the short direction.

  $\Longrightarrow$ it is called a "one-way slab".

8.2 Design Methods:

* Since the loading on a one-way slab is nearly all transferred in the short direction, such a slab continuous over several supports may be treated as a beam.

* ACI Code/Section 8.3.3 permits the use of moment and shear coefficients in the case of two or more approximately equal spans (the larger of two adjacent spans not exceeding the shorter by more than 20%) with loads uniformly distributed, where the unit live load does not exceed three times the unit dead load.

* these coefficients are in terms of clear span L_n and the values are given for critical locations, that is, faces of support for shears — and negative moments, and midspan regions for positive moments.

* When conditions of ACI Code/Section 8.3.3 are <u>not</u> satisfied, an elastic analysis is required.

* Approximate methods are permitted in "buildings of usual types of construction, spans, and story heights."

* See the _PCA publication_ "_Continuity in Concrete Building Frames_."

* For an elastic analysis, negative bending moments are obtained at centers of supports. To get the negative bending moment at the face of support, we reduce the maximum value at the center of support by a quantity equal to $\frac{1}{3}Vb$, where V is the shear force at either the center or at the face of support, and b is the width of support.

---

* Suppose AB is fixed at A and B.

$$M_A = -\frac{wL^2}{12}.$$

$$M_{A'} = -\frac{wL^2}{12} + \frac{wL}{2}\left(\frac{1}{2}b\right)$$

$$-w\left(\frac{b}{2}\right)\cdot\frac{1}{2}\left(\frac{b}{2}\right)$$

$$\therefore M_{A'} = -\frac{wL^2}{12} + \frac{wLb}{4} - \frac{wb^2}{8}$$

$$\therefore M_{A'} = \boxed{-\left(\frac{wL^2}{12} + \frac{wLb}{4} + \frac{wb^2}{8}\right)}$$

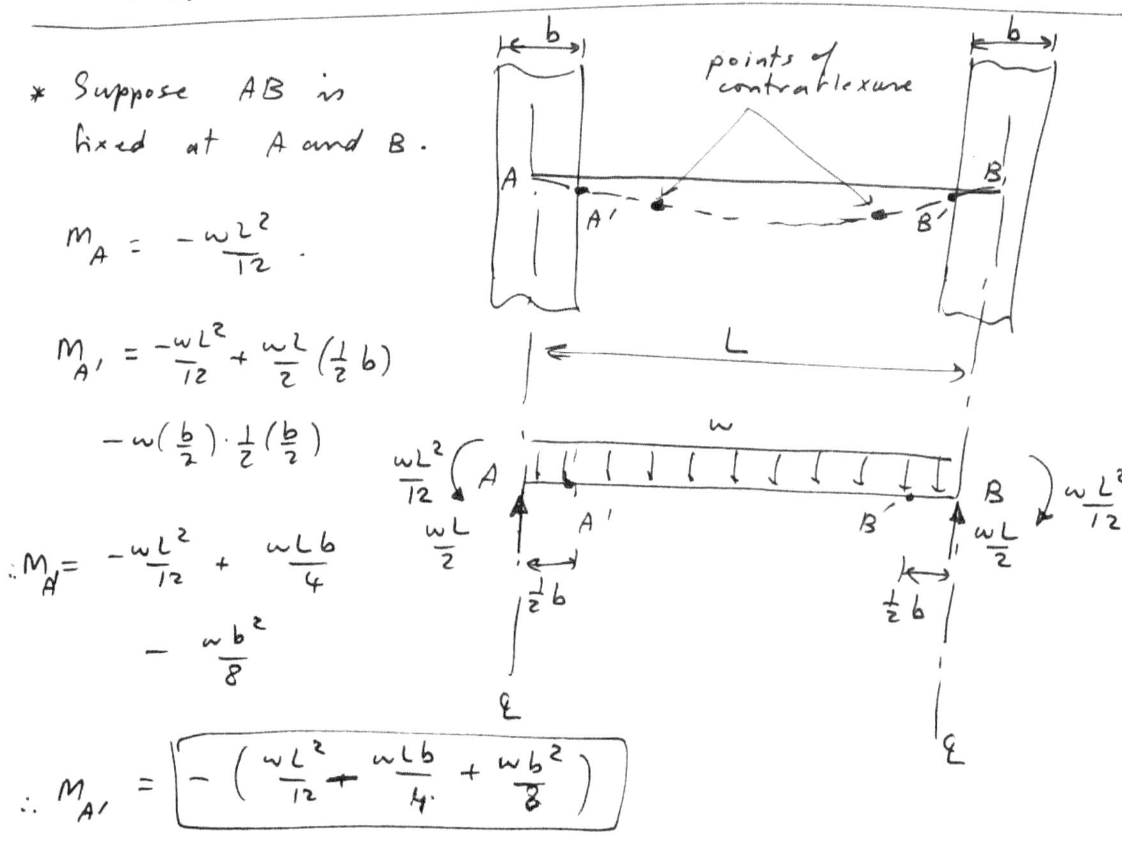

Suppose AB is fixed at A' and B':

$$M_{A'} = -\frac{w}{12}(L-b)^2 = \boxed{-\left(\frac{wL^2}{12} - \frac{wLb}{6} + \frac{wb^2}{12}\right)}$$

If the term involving b^2 is neglected:

$$M_A = -\left(\frac{wl^2}{12} - \frac{wlb}{4}\right)$$

$$M_{A'} = -\left(\frac{wl^2}{12} - \frac{wlb}{6}\right)$$

$$M_{A'} - M_A = -\left(\frac{wl^2}{12} - \frac{wlb}{6}\right) + \left(\frac{wl^2}{12} - \frac{wlb}{4}\right)$$

$$= \frac{wlb}{6} - \frac{wlb}{4}$$

$$= \frac{2wlb - 3wlb}{12}$$

$$= -\frac{wlb}{12}$$

$\Longrightarrow \quad M_A > M_{A'} \quad$ by the amount $\frac{wlb}{12}$.

$\therefore \quad M_A - M_{A'} = \frac{wlb}{12} \quad$ (difference in moment at center of support and moment at face of support). (Both assumed fixed).

$\left[\text{but} \quad M_{A'} = -\left(\frac{wl^2}{12} - \frac{wlb}{6}\right) \text{ is } \underline{\text{more}} \text{ realistic} \right.$

since for a relatively stiff support, such as a column or girder, there can be very little change in rotation or deflection between points A and A'.

$\hookrightarrow$ Moment at Face of Support is less than Moment at Center of Support by $\frac{wlb}{6} = \left(\frac{1}{2}wl\right)\frac{b}{3} = \frac{Vb}{3}$.

$$\left(\text{since } V = \tfrac{1}{2}wl\right).$$

8.3 Thickness of Slab:

* In designing a one-way slab, a typical imaginary strip 1-m wide is usually considered.

* the continuous slab may then be designed as a continuous beams having a known width of 1m; the slab thickness is the only unknown.

* the thickness of the slab depends on the deflection, bending, and shear requirements.

* Deflection requirements are imposed to prevent excessive deformations that might adversely affect the serviceability of the structure.

* According to ACI-Table 9.5a, one-way slabs have at least a minimum slab thickness (for Grade 60 steel) of $\frac{L}{20}$, $\frac{L}{24}$, $\frac{L}{28}$, or $\frac{L}{10}$, depending on whether L is the length of a simply-supported, a one-end continuous, a both ends continuous, or a cantilever span.

* If the slab supports or is attached to construction likely to be damaged by large deflections, deflection must be computed and shown to satisfy the limits of ACI-Table 9.5b.

* Design one-way slabs according to the Strength Design Method (see Chapter 3).

* In continuous (equal) spans, the negative moment at the exterior face of the first interior support is the largest; therefore, this negative moment should be used to establish the slab thickness.

* the shear requirement does not usually control, but it must be checked.

* Because of practical space limitations, shear reinforcement is not used in a slab; thus $V_u \leq 0.5 \phi V_c$ (at distance d from face of support).

* ACI Code/Section 7.7.1-c specifies that the concrete protective covering for reinforcement ($\phi 36$ mm and smaller) in slabs shall be not less than 20 mm at surfaces not exposed directly to the ground or weather.

* When the top of a monolithic slab is the wearing surface and when unusual wear is expected as in warehouses or industrial buildings, it has been customary

to use an additional depth of $\frac{1}{2}$ in. (≈ 12.5 mm) of concrete protective covering over that required by the design of the member.

Example 1 :

Establish the thickness of the floor slab shown in the second-floor framing plan (see the figure) for a service live load of 4.8 kN/m². Use $f_c' = 20$ MPa, $f_y = 275$ MPa, and the strength method of the ACI Code. Assume an exterior staircase so that no openings are to be made in the slabs. The figure shows a section through slabs 2S1 – 2S2 – 2S3.

Solution :

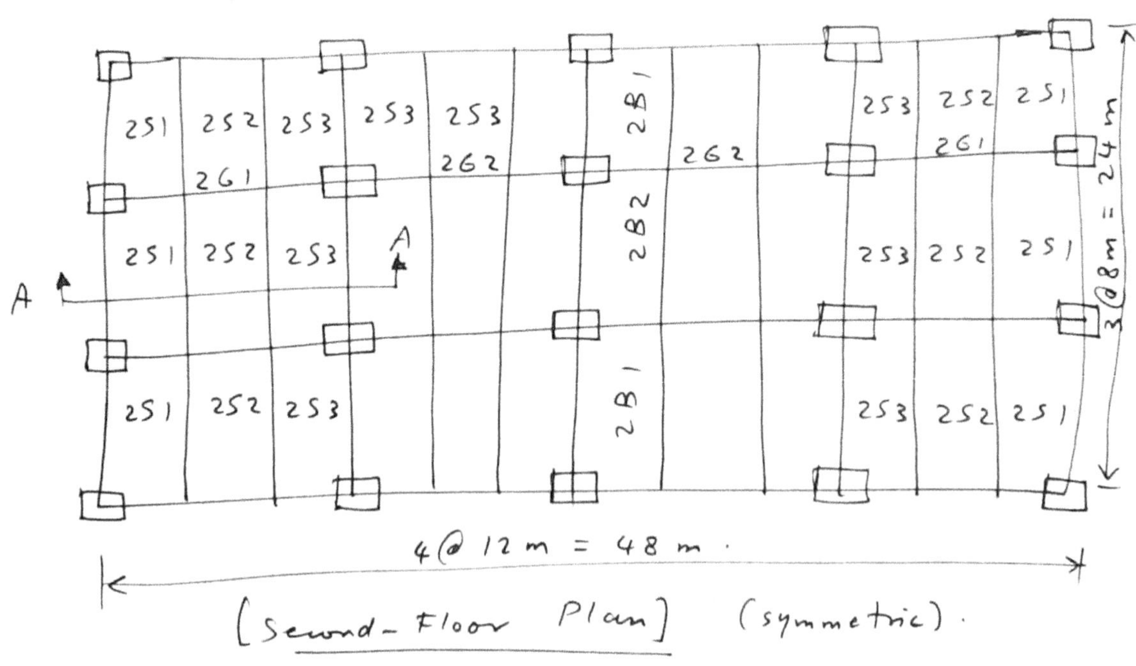

[Second - Floor Plan]   (symmetric).

Use the ACI moment coefficients and their corresponding shear and moment diagrams (assume that the live load does not exceed three times the dead load).

(a) <u>Minimum Thickness</u> : (to check deflections):

From Table 9.5a (ACI), For spans with one end and both ends continuous, the respective minimum thicknesses h are $\frac{L}{24}$ and $\frac{L}{28}$, $\qquad \underline{L = 4\,m}$.

these values are for $f_y = 400\,MPa$.

For $f_y \neq 400\,MPa$, multiply the values by $\left(0.4 + \frac{f_y}{700}\right)$

$$\therefore \text{minimum } h = \frac{L}{24}\left(0.4 + \frac{f_y}{700}\right) = \frac{L}{24}\left(0.4 + \frac{275}{700}\right)$$

$$= \frac{L}{24}(0.793) \simeq \frac{L}{30} = \frac{4 \times 1000}{30}$$

$$= 133.33 \text{ mm (For } 2S1).$$

$$\text{minimum } h = \frac{L}{28}(0.793) \simeq \frac{L}{35} = \frac{4 \times 1000}{35} = 114.3 \text{ mm}$$

$$\text{(for } 2S2 \text{ and } 2S3).$$

Assume $h = 120\,mm$.

Weight of slab $= \frac{120}{1000} \times 25 = 3 \text{ kN/m}^2$.

(b) Bending Moment Requirements:

Assume width of supporting beam $= 330\,mm$.

$$w_u = 1.4\,w_D + 1.7\,w_L$$

$$= 1.4(3) + 1.7(4.8) = 12.36 \text{ kN/m}^2$$

$$\equiv 12.36 \text{ kN/m per meter of width}$$

$$L_n = \text{clear span} = 4 - \frac{330}{1000} = 3.67 \text{ m}.$$

$$M_u = \frac{1}{10} w_u L_n^2 = \frac{1}{10}(12.36)(3.67)^2 = 16.65 \text{ kN}\cdot\text{m per meter of width}.$$

Choose $\rho = \frac{1}{2}\rho_{max} = 0.375\,\rho_b$.

$$\beta_1 = 0.85$$

$$\rho_b = \frac{0.85\, f_c'}{f_y}\,\beta_1 \left(\frac{600}{600 + f_y}\right)$$

$$= \frac{0.85(20)(0.85)}{275}\left(\frac{600}{600 + 275}\right) = 0.03603$$

Take $\rho = 0.375\,(0.03603) = 0.01351 = 1.351\%$.

$$m = \frac{f_y}{0.85 f_c'} = \frac{275}{0.85\,(20)} = 16.18 .$$

$$R_n = \rho f_y \left(1 - \tfrac{1}{2}\rho m\right)$$

$$= (0.01351)(275)\left[1 - \tfrac{1}{2}(0.01351)(16.18)\right]$$

$$= 3.309 \;\; MPa$$

required $M_n = \dfrac{M_u}{\phi} = \dfrac{16.65}{0.90} = 18.5$ kN·m per meter of width.

$$\text{required } bd^2 = \frac{\text{required } M_n}{R_n} = \frac{18.5}{3.309 \times 1000}$$

$$= 0.0055908 \;\; m^3$$

$$= 5.5908 \times 10^{6} \;\; mm^3 .$$

but $\quad b = 1\,m = 1000\,mm$.

$$\therefore \text{ required } d = \sqrt{\frac{5.5908 \times 10^6}{b}} = \sqrt{\frac{5.5908 \times 10^6}{1000}} = 75\,mm.$$

Assume $\phi 16\,mm$ bars $(d_b = 16\,mm)$ and $20\,mm$ cover,

required $h = 75 + \dfrac{16}{2} + 20 = 103\,mm$.

$\therefore$ Use $h = 120\,mm$ as originally done.

Provided $d = 120 - \dfrac{16}{2} - 20 = 92\,mm$.

(c) <u>Shear Requirement:</u>

From the ACI moment and shear diagrams,

$$\text{Max. } V_u = 0.5583\, w_u L_n = 0.5583\,(12.36)(3.67)$$

$$= 25.32 \;\; kN \text{ per meter of width.}$$

$$\phi V_c = \phi\left(\tfrac{1}{6}\sqrt{f_c'}\, b_w\, d\right) = 0.85\left[\tfrac{1}{6}\sqrt{20}\,\frac{(1000)(92)}{1000}\right] = 58.3 \;\; kN .$$
$$\text{(simplified method).}$$

$$0.5 \phi V_c = \frac{58.3}{2} = 29.1 \text{ kN}$$

$$\Rightarrow \quad V_u = 25.32 \text{ kN} < 0.5 \phi V_c = 29.1 \text{ kN}$$

$$\Rightarrow \quad \text{No shear reinforcement is required.}$$

<u>Note</u> : V_u should be calculated at a distance d from face of support. But in this case, the factored shear would have been even less than 25.32 kN.

8.4 Choice of Reinforcement:

* Choice of reinforcement depends on:
  1. steel area
  2. development length calculations.

* Compute the steel areas at the principal sections (i.e. at the middle and at the ends of each span).

* Make a <u>tentative</u> choice of reinforcement.

* One possible arrangement of reinforcement, though <u>not</u> commonly used, is shown in the figure below. (straight bars and bent (trussed) bars are placed alternately).

* The straight bars across the bottom are ordinarily one size smaller than the bent bars.

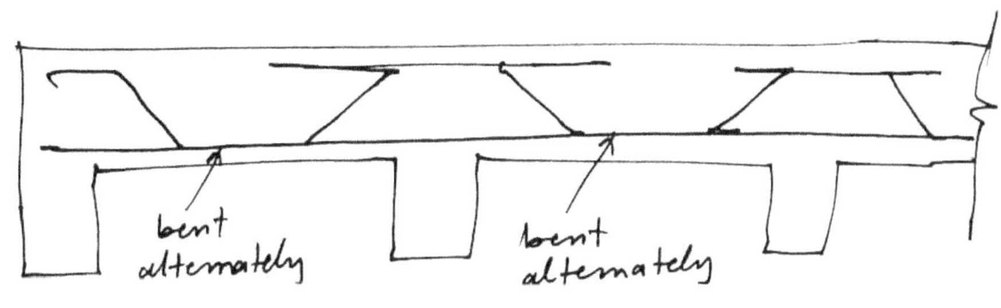

* Frequently, in thin slabs of under 125 mm of thickness, straight bars are used in both the top and bottom in all spans. Because of economics, most designers use <u>straight bars in all cases.</u>

* Development length requirements should next be examined.

* The positive moment bars that are bent up into the negative moment region have ample embedment.

* the bars that continue along the bottom of the slab and extend into the support at least 150 mm (ACI Code/Section 12.11.1) must be checked for adequate development at points of inflection (ACI Code/Section 12.11.3)

* In addition, the bars extending straight across the top beyond the bend-down points must be checked for satisfactory embedment (ACI Code/Section 12.12.3).

---

* Other Limitations for Slabs:

(1) In structural slabs of uniform thickness (ACI Code/10.5.3), the minimum amount of reinforcement in the direction of the span shall $\underline{not}$ be less than that required for shrinkage and temperature reinforcement. (ACI Code/Section 7.12).

(2) The principle reinforcement shall be centered $\underline{not}$ farther apart than three times the span thickness, $\underline{nor}$ more than 450 mm (ACI Code/Section 7.6.5).

Example 2 :

Choose the arrangement of reinforcement in the 120-mm floor slab 2S1, 2S2, and 2S3, as designed in Example 1.

Solution :

(a) Area Requirements:

See the table on the next page.

Reinforcement for One-Way Slab of Example -1- , Using ACI Moment Coefficients

	2S1			2S2			2S3		
	Support	Middle	Support	Support	Middle	Support	Support	Middle	Support
ACI moment Coefficient (α)	$-\frac{1}{24}$	$+\frac{1}{14}$	$-\frac{1}{10}$	$-\frac{1}{11}$	$+\frac{1}{16}$	$-\frac{1}{11}$	$-\frac{1}{11}$	$+\frac{1}{16}$	$-\frac{1}{11}$
$M_u = \alpha w_u L_n^2$ (kN·m) $= \alpha (12.36)(3.67)^2$	-6.94	$+11.89$	-16.65	-15.13	$+10.40$	-15.13	-15.13	$+10.40$	-15.13
$R_n = \dfrac{M_u}{\Phi b d^2}$ (MPa)	0.911	1.56	2.19	1.986	1.365	1.986	1.986	1.365	1.986
$m = f_y/(0.85 f_c')$ $\rho = \frac{1}{m}\left(1 - \sqrt{1 - \frac{2mR_n}{f_y}}\right)$ $(m = 16.18)$ $\left(\text{minimum } \rho = \frac{1.4}{f_y} = 0.0051\right)$	$\boxed{0.0034}$	0.00596	0.00856	0.00770	0.00518	0.00770	0.00770	0.00518	0.00770
Required A_s (mm²) $= \rho b d$ $= \rho (1000)(92)$	469	548	788	708	477	708	708	477	708
Provided A_s (mm²)	Ø12mm @ 240mm (st) (471 mm²)	(798 mm²) Ø12mm@240mm (bt) Ø10mm@240mm (st)	Ø12mm @ 240mm (bt) Ø12mm @ 240mm (bt) (942 mm²)	Ø12mm @ 240mm (bt) Ø12mm @ 240mm (bt) (942 mm²)	(798 mm²) Ø12mm@240mm (bt) Ø10mm @ 240mm (st)	Ø12mm @ 240mm (bt) Ø12mm @ 240mm (bt) (942 mm²)	Ø12mm @ 240mm (bt) Ø12mm @ 240mm (bt) (942 mm²)	(798 mm²) Ø12mm @ 240mm (bt) Ø10mm @ 240mm (st)	Ø12mm @ 240mm (bt) Ø12mm @ 240mm (bt) (942 mm²)

Ø12mm @ 120mm

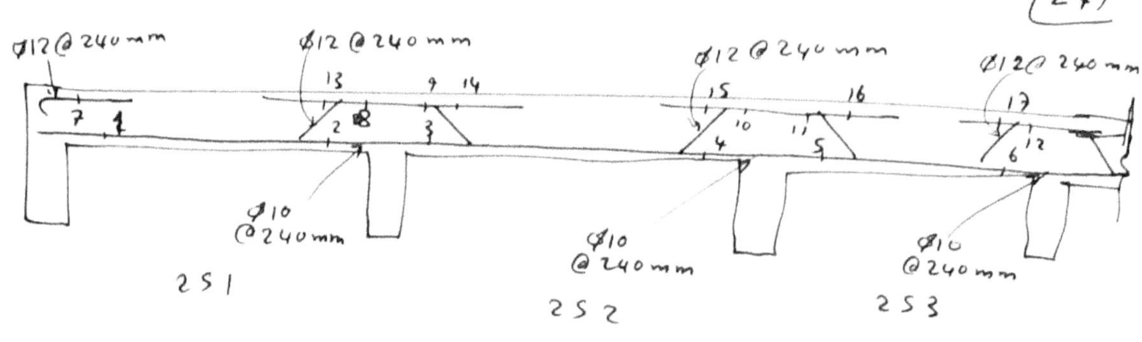

(b) Development Length Requirements at Inflection Points:

(1) Check development lengths at points 1 and 2 − 6.

(2) Anchorage length must be provided in both directions from the maximum moment points at the faces of support (points 7 − 12).

(3) Anchorage of the straight portion of the bars at the top of the slab beyond extreme points of inflection (points 13 − 17).

From the ACI Shear and Moment Diagrams, the shears at inflection points on the typical 1-m width of slab are:

$$V_1 = V_2 = 0.3780 \, w_u L_n = 0.3780 \,(12.36)\,(3.67)$$
$$= 17.15 \text{ kN}.$$

$$V_3 = V_4 = V_5 = V_6 = 0.3535 \, w_u L_n = 0.3535 \,(12.36)\,(3.67)$$
$$= 16.04 \text{ kN}.$$

For the bars (∅10 mm) that extend past the inflection points into the supports,

$$\frac{l_d}{d_b} = \frac{f_y \, \alpha \, \beta \, \lambda}{2 \sqrt{f_c{}'}} \qquad (\text{Table } 6.2)$$

$$= \frac{(275)\,(1.0)\,(1.0)\,(0.8)}{2\sqrt{20}} = 25.0$$

$$l_d = 25.0 \, d_b = 25.0 \,(10) = 250 \text{ mm} < 300 \text{ mm}.$$

Note: l_d must <u>not</u> be less than 300 mm.

$$\Rightarrow \text{Use } l_d = 300 \text{ mm}.$$

The requirement of ACI/Code/Section 12.11.3 at inflection points (such as point 2) is:

$$l_d \leq \frac{M_n}{V_u} + l_a \qquad \text{(at point of inflection)}$$

For $\phi 10\,mm$ @ $240\,mm$, $A_s = 327\,mm^2$.

$$C = 0.85 f_c' \, ab = 0.85 \times (20 \times 1000) \times a \times 1 \qquad b = 1\,m$$
$$= 17,000\,a$$

$$T = f_y A_s = (275 \times 1000)(327 \times 10^{-6})$$

$$C = T \implies a = \frac{(275 \times 1000)(399 \times 10^{-6})}{17,000} = 0.00645\,m$$
$$= 6.45\,mm.$$

$$M_n = T\left(d - \frac{a}{2}\right) = (275 \times 1000)(327 \times 10^{-6})\left[\frac{92}{1000} - \frac{6.45}{2(1000)}\right]$$

$$= 9.74\ kN \cdot m.$$

$$V_u = 17.15\ kN.$$

$$l_a = \begin{cases} 12 d_b = 12(10) = 120\,mm. \\ d = 92\,mm. \end{cases} \text{take } l_a = 120\,mm.$$

$$\therefore \quad \frac{M_n}{V_u} + l_a = \left(\frac{9.74}{17.15}\right)^{\times 1000} + 120 = 688\,mm.$$

$$\therefore \quad l_d = 300\,mm \quad < 688\,mm. \qquad \underline{o.k.}$$

At the exterior end, (simple support),

$$l_d \leq 1.3\frac{M_n}{V_u} + l_a \qquad (\text{is also satisfied})$$

(c) Cutoff Points For Negative Moment Reinforcement: 243

For $\phi 12\,mm$ @ $240\,mm$, $A_s = 471\,mm^2$.

$$\frac{l_d}{d_b} = \frac{f_y\,\alpha\,\beta\,\lambda}{2\sqrt{f_c'}} = \frac{(275)(1.0)(1.0)(0.8)}{2\sqrt{20}} = 24.6$$

$$l_d = 24.6\,d_b = 24.6\,(12) = 295\,mm \quad < \quad 300\,mm$$

$$\implies \quad Use \quad l_d = 300\,mm.$$

<u>Note</u> : Even though the $\phi 12\,mm$ bars are in the top of the slab, they are <u>not</u> top bars because there is less than $300\,mm$ of concrete cast beneath them.

To satisfy ACI 12.12.3, find the distance to the point of inflection.

$$distance \approx 0.3\,L_n = 0.3\,(3.67) = 1.10\,m.$$

(d) <u>Bent-up or Bend-down Location</u> :

Since the bend slope is usually about $45°$, the bend can be defined by locating either the bend-up or the bend-down location.

The ACI Moment Diagram (Figure 7.5.2 (a)) can be used to determine the bend-up location near the right end of slab 251.

Moment corresponding to $\phi 10\,mm$ @ $240\,mm$,

$$\left(\frac{327}{798}\right) * \left(\frac{1}{14}\,w_u\,L_n^2\right) = 0.02927\,w_u\,L_n^2 .$$

$$\left(\frac{Area\ of\ \phi 10\,mm\ @ 240\,mm}{Total\ A_s}\right) * \frac{1}{14}\,w_u\,L_n^2 .$$

Distance to Bend-up location $= 0.5244 L_n - L_n \sqrt{(\frac{1}{14} - 0.02927)(2)}$

$$= 0.234 L_n$$

(see Figure 7.5.2(2)).

* this distance $\approx 0.25 L_n$ (from face of support to the bend-up location).

8.5   Continuity Analysis.

* When the necessary conditions for using the ACI Moment Coefficients and shear coefficients are <u>not</u> satisfied, an elastic analysis is required.
* In this case, you can use a computer program.

8.6 Shrinkage and Temperature Reinforcement:

* Reinforcement for shrinkage and temperature stresses normal to the main reinforcement is required in structural floors and roof slabs, where the main reinforcement extends in one direction only (ACI Code/ Section 7.12.2).

* Such reinforcement shall provide (ACI Code/Section 7.12.2) for the following minimum ratios of reinforcement area to gross concrete area, but in no case shall such reinforcing bars be placed farther apart than <u>five</u> times the slab thickness nor more than 450 mm (ACI Code/Section 7.12.2.2).

Type of Slab	ρ
1. Slabs where Grades 40 or 50 deformed bars are used $(275 \rightarrow 345 \text{ MPa})$.	0.0020
2. Slabs where Grade 60 deformed bars are used (400 MPa)	0.0018
3. Slabs where reinforcement with yield strength exceeding 400 MPa measured at a yield strain of 0.35% is used	$0.0018 \left(\dfrac{400}{f_y}\right)$

Example 3:

Design the shrinkage and temperature reinforcement in the floor slab of Examples 1 and 2.

Solution:

Area of shrinkage and steel reinforcement is:

$$A_s = \rho b h = (0.0020)(1000)(120) = 240 \text{ mm}^2.$$

Use $\phi 10 \text{ mm}$ @ 300 mm c/c $\implies A_s = 262 \text{ mm}^2 > 240 \text{ mm}^2$